Inside the black box

Technology and economics

Inside the black box

Technology and economics

NATHAN ROSENBERG
Professor of Economics, Stanford University

Published by the Press Syndicate of the University of Cambridge
The Pitt Building, Trumpington Street, Cambridge CB2 1RP
40 West 20th Street, New York, NY 10011-4211, USA
10 Stamford Road, Oakleigh, Melbourne 3166, Australia

First published 1982
Reprinted 1983, 1984, 1986, 1988, 1990, 1993, 1995

Printed in the United States of America

Library of Congress Cataloging-in-Publication Data

Rosenberg, Nathan, 1927–
Inside the black box.
Includes index.
1. Technological innovationd.
2. Technology – Social aspects. 3. Economic
development. I. Title.
HC79.T4R673 1982 338'.06 82–4563

ISBN 0-521-24808-6 hardback
ISBN 0-521-27367-6 paperback

Contents

Preface

The central purpose of this book may be simply stated. Economists have long treated technological phenomena as events transpiring inside a black box. They have of course recognized that these events have significant economic consequences, and they have in fact devoted considerable effort and ingenuity to tracing, and even measuring, some of these consequences. Nevertheless, the economics profession has adhered rather strictly to a self-imposed ordinance not to inquire too seriously into what transpires inside that box.

The purpose of this book is to break open and to examine the contents of the black box into which technological change has been consigned by economists. I believe that by so doing a number of important economic problems can be powerfully illuminated. This is because the specific characteristics of certain technologies have ramifications for economic phenomena that cannot be understood without a close examination of these characteristics. Thus, I attempt to show in the following pages how specific features of individual technologies have shaped a number of developments of great concern to economists: the rate of productivity improvement, the nature of the learning process underlying technological change itself, the speed of technology transfer, and the effectiveness of government policies that are intended to influence technologies in particular ways.

The separate chapters of this book reflect a primary concern with some of the distinctive aspects of industrial technologies in the twentieth century: the increasing reliance upon science, but also the considerable subtlety and complexity of the dialectic between science and technology; the rapid growth in the development costs associated with new technologies, and the closely associated phenomena of lengthy lead times and the high degree of technological uncertainty associated with precisely predicting the eventual performance characteristics of newly emerging technologies; the changing structure of interindustry relationships, such as that between the makers of capital goods and their eventual users; and the changing characteristics of a technology over the course of its own life cycle. Each of the chapters in Part II represents an attempt to identify some significant characteristics of specific advanced industrial technologies – or of the process by which such technologies have emerged and have been introduced into the economy. The chapters in Parts III and IV continue this examination against the backdrop of a concern with issues of public policy and with the implications of technology transfer in the international context.

The book opens with a broad survey, in Part I, of the historical literature on technical change. It attempts to provide a guide to a wide range of writings, including those by some social historians and social theorists as well as economic historians and economists, that illuminate technological change as a historical phenomenon. It should not be necessary to belabor two points: (1) that past history is an indispensable source of information to anyone interested in characterizing technologies, and (2) that both the determinants and the consequences of technological innovation raise issues that go far beyond the generally recognized domain of the economist and the economic historian. The first chapter discusses aspects of the conceptualization of technological change and then goes on to consider what the literature has had to say on (1) the rate of technological change, (2) the forces influencing its direction, (3) the speed with which new technologies have diffused, and (4) the impact of technological change on the growth in productivity.

A separate chapter is devoted to Marx. Marx's intellectual impact has been so pervasive as to rank him as a major social force *in* history as well as an armchair interpreter *of* history. And yet, curiously enough, I argue that Marx's analysis of technological change opened doors to the study of the technological realm through which hardly anyone has subsequently passed.

Part II is, in important respects, the core of the book. Each of its chapters advances an argument about some significant characteristics of industrial technologies, characteristics that are typically suppressed in discussions of technological change conducted at high levels of aggregation or lacking in historical specificity. Chapter 3 explores a variety of less visible forms in which technological improvements enter the economy. Each of these forms, it is argued, is important in determining the connections between technological innovations and the growth of productivity flowing from innovation. Chapter 4 explicitly considers some significant characteristics of different energy forms. It became a common practice in the 1970s, following the Arab oil embargo, to treat energy as some undifferentiated mass expressible in Btus which it was in society's interests to minimize. This chapter examines some of the complexities of the long-term interactions between technological change and energy resources. It emphasizes, in particular, the frequently imperfect substitutability among energy sources in industrial contexts and the consequent suboptimality of criteria for energy utilization that fail to take specific characteristics of different energy forms into account.

Chapter 5, "On Technological Expectations," addresses an issue that is simultaneously relevant to a wide range of industries – indeed, to all industries that are experiencing, or are expected to experience, substantial rates of technical improvement. I argue that rational decision making with respect to the adoption of an innovation requires careful consideration of prospective rates of technological innovation. Such a consideration will often lead to counterintuitive decisions, including slow adoption rates that, from other perspectives, may appear to be irrational. Expectations about the future behavior of technological systems and their components are shown to be a major and neglected factor in the diffusion of new technologies.

The last two chapters of Part II are primarily concerned with issues of greatest relevance to high-technology industries – industries in which new product development involves large development costs, long lead times, and considerable technological uncertainty (especially concerning product performance characteristics) and that rely in significant ways upon knowledge that is close to the frontiers of present-day scientific research. Chapter 6, "Learning by Using," identifies an important source of learning that grows out of actual experience in using products characterized by a high degree of system complexity. In contrast to learning by doing, which deals with skill improvements that grow out of the productive process, learning by using involves an experience that begins where learning by doing ends. The importance of learning by using is explored in some detail with respect to aircraft, but reasons are advanced suggesting that it may be a much more pervasive phenomenon in high-technology industries.

The final chapter in Part II, "How Exogenous Is Science?" looks explicitly at the nature of science–technology interactions in high-technology industries. It examines some of the specific ways in which these industries have been drawing upon the expanding pool of scientific knowledge and techniques. The chapter also considers, however, a range of much broader questions concerning the institutionalization of science and the manner in which the agenda of science is formulated in advanced industrial societies. Thus, a major theme of the chapter is that, far from being exogenous forces to the economic arena, the content and direction of the scientific enterprise are heavily shaped by technological considerations that are, in turn, deeply embedded in the structure of industrial societies.

The three chapters constituting Part III share a common concern with the role of market forces in shaping both the rate and the direction of innovative activities. They attempt to look into the composition of forces constituting the demand and the supply for new products and processes, especially in high-technology industries. This analysis, in turn, has direct implications for government concern with accelerating the rate of innovative activity. Thus, policy considerations emerge as an important element of these chapters.

Chapter 8 examines the history of technical change in the commercial aircraft industry over the fifty-year period 1925–75. This industry has been, and remains, a remarkable success story in terms of both productivity growth and continued American success in international markets. For a variety of reasons, including the strategic military importance of aircraft and a concern with passenger safety, the federal government's role has been particularly prominent with respect to aircraft. This chapter evaluates the impact of government policies and considers the possible relevance of these policies to other industries. Chapter 9 examines the ongoing technological revolution embodied in very-large-scale integration. It points out that there are a variety of mediating factors that stand between an expanding technological capability and commercial success. The growth in circuit-element density, with the resulting dramatic improvement in the capability of a single chip, offers a great potential for the application of electronic techniques in many fields. The success of such applications will turn upon developments internal to the industry, but also upon the creation of mechanisms that will translate this new technological capability into tangible economic advantages. Chapter 10 focuses not upon an individual industry but upon a number of recent empirical studies of technical change. These studies, which share an emphasis upon the dominant role of market demand in the innovation process, have been widely cited as providing an adequate basis for a successful government innovation policy. It is argued that these studies are, analytically and conceptually, seriously incomplete. The chapter attempts to provide a more comprehensive framework for both analysis and policy formulation.

Finally, the two chapters of Part IV place the discussion of technological change in an international context, with the first chapter oriented toward its long history and the second toward the present and the future. Chapter 11 pays primary attention to the transfer of industrial technology from Britain to the rest of the world. This transfer encompasses a large part of the story of worldwide industrialization, because nineteenth-century industrialization was, in considerable measure, the story of the overseas transfer of the technologies already developed by the first industrial society. Particular attention is devoted to the conditions that shaped the success of these transfers, but a central concern is their eventual impact upon the technology-exporting country. The last chapter speculates about the prospects for the future from an American perspective, a perspective that is often dominated by apprehension over the loss of American technological leadership, especially in high-technology industries. By drawing upon some of the distinctive characteristics of high-technology industries, an attempt is made to identify possible elements of a future scenario. I am confident that the world economy of the 1990s will be powerfully shaped by the international distribution of technological capabilities; but it will also be shaped by economic and social forces that strongly influence the comparative effectiveness with which the available technologies are exploited. I also suspect that the world of the 1990s will be a good deal more complex – and more interesting – than the one currently depicted by the harbingers of The Japanese Challenge, just as the scenario presented in Jean-Jacques Servan-Schreiber's *The American Challenge,* published in 1968, bore little resemblance to the subsequent decade of the 1970s.

This book is, in many respects, a continuation of the intellectual enterprise that was embodied in my earlier book, *Perspectives on Technology.* Whereas in the introduction to that book I stated that my interest in coming to grips with technological change had had the effect of transforming an economist into an economic historian, I am now inclined to say that much of the content of the present book can be read as the musings of an economic historian who has stumbled – not entirely by accident! – into the twentieth century. For the benefit of economic historians who still think of themselves as young, and who take it for granted that to study history is to study some remote past, I must point out that the twentieth century is, by now, mostly history.

PART I

Views of technical progress

1 The historiography of technical progress

To encompass the entire historiography of technical progress in one essay is impossible, even if the essay were allowed to grow far longer than the present one. For, in a fundamental sense, the history of technical progress is inseparable from the history of civilization itself, dealing as it does with human efforts to raise productivity under an extremely diverse range of environmental conditions. Even if we were to define technology in a relatively narrow "hardware" sense – which we will not – and to exclude organizational, institutional, and managerial factors, the range of materials that one might wish to mention would still be disconcertingly large. What follows, therefore, is necessarily highly selective.

This essay will first consider the nature and character of technical progress. Successive sections will then explore the most relevant literature on the rate of technical progress, the direction of technical progress, the diffusion of new technologies, and finally, the impact of technical progress upon productivity growth.

Definition and characterization of technical progress

A central problem in examining technical progress, and one that makes it difficult even to define or characterize readily, is that it takes many different forms. For technical progress is not one thing; it is many things. Perhaps the most useful common denominator underlying its multitude of forms is that it constitutes certain kinds of knowledge that make it possible to produce (1) a greater volume of output or (2) a qualitatively superior output from a given amount of resources.

The second category is most important and should not be regarded as a minor afterthought. The great bulk of the writing by economists on

This paper was originally published in Italian under the title "Progresso Tecnico: L'Analisi Storica," in *Il Mondo Contemporaneo, vol. VIII: Economia E Storia* – 2, 1978, pp. 626–45. The English version is reprinted by permission of the editors of *Dizionario critico di storia contemporanea*.

the subject of technical change – both theoretical and empirical – treats the phenomenon as if it were solely cost-reducing in nature, that is, as if one could exhaust everything of significance about technical change in terms of the increases in output per unit of input that flow from it. Technical progress is typically treated as the introduction of new processes that reduce the cost of producing an essentially unchanged product. Perhaps the main reasons for the popularity of this approach are these: It is a useful simplification that makes it possible to analyze a wide range of problems with a relatively simple analytical apparatus, and it allows a quantitative approach to innumerable interesting economic questions. At the same time, however, to ignore product innovation and qualitative improvements in products is to ignore what may very well have been the most important long-term contribution of technical progress to human welfare. Western industrial societies today enjoy a higher level of material welfare not merely because they consume larger per capita amounts of the goods available, say, at the end of the Napoleonic wars. Rather, they have available entirely new forms of rapid transportation, instant communication, powerful energy sources, life-saving and pain-reducing medications, and a bewildering array of entirely new goods that were undreamed of 150 or 200 years ago. To exclude product innovation from technical progress, especially when we are considering long historical periods, is to play Hamlet without the prince.

Of course, not all economists have ignored product innovation. Not surprisingly, the subject has been treated most carefully and imaginatively by economists who have also been serious students of economic history. To begin with, Simon Kuznets has pointed out that whether an innovation concerns a product or a process depends very much upon whose perspective one is adopting (Kuznets, 1972). Process innovations typically involve new machinery or equipment in which they are embodied; this machinery or equipment constitutes a product innovation from the point of view of the firm that produces it. Thus, the Bessemer converter was a process innovation to iron and steel manufacturers but a product innovation to the suppliers of equipment to the iron and steel industry. Furthermore, Kuznets had richly documented as long ago as 1930 (Kuznets, 1930) the central role of product innovation in long-term economic growth.[1] Kuznets argued that high aggregative growth rates in industrial economies have reflected continuous shifts in product and industry mix. All rapidly growing industries eventually experience a

[1] An earlier summary of the book had appeared in the *Journal of Economic and Business History,* August 1929, pp. 534–60, and is reprinted in Kuznets (1953), chap. 9, "Retardation and Industrial Growth." See also Kuznets (1971), chap. 7.

slowdown in growth as the cost-reducing impact of technical innovation diminishes. Furthermore, because of the typically low long-term income and price elasticity of demand for old consumer goods, further cost-reducing innovations in these industries will have a relatively small aggregative impact. Therefore, continued rapid growth requires the development of new products and new industries.

Of course, Kuznets has not been entirely alone in his emphasis upon the importance of new products. Joseph Schumpeter had emphasized throughout his life the central role of technical progress in understanding the dynamics of capitalist growth. His great work, *Business Cycles* (1939), focused powerfully upon the historical role of technological innovation in accounting for the high degree of instability in capitalist economies. His later book, *Capitalism, Socialism, and Democracy* (1942), is a virtual paean to the beneficent impact of what he called the "perennial gales of creative destruction." These "gales" were closely tied to product innovation that swept away old industries producing old products. Thus, economic progress, for Schumpeter, did not consist of price cutting among harness makers. The competitive behavior that really mattered in the long run came from the innovative acts of automobile manufacturers, which abolished harness making as an economic activity. Thus, for Schumpeter, product innovation had fundamental implications for understanding the nature of capitalism as a historical force as well as the nature of the competitive process. For economists had erroneously assumed that the problem "is how capitalism administers existing structures, whereas the relevant problem is how it creates and destroys them" (Schumpeter, 1942, p. 84).

Schumpeter has also profoundly influenced the approach of economists and economic historians to the study of technical progress by his stress upon its discontinuous nature. To begin with, he defined innovation very broadly as a shift in a production function that might have a variety of causes. These causes encompass much more than technical progress in a narrow sense – that is, product or process innovation. In addition, they may include the opening up of a new market, the acquisition of a new source of raw materials, or a structural reorganization of an industry (Schumpeter, 1934, p. 66). Of even greater importance for our present discussion is Schumpeter's great emphasis upon technical progress as constituting major breaks, giant discontinuities with or disruptions of the past. It was an emphasis that fitted particularly well both with his analysis of the sociology of capitalist society[2] and with his search for the strategic factor in business cycles. (The clustering of

[2] Schumpeter states that "successful innovation . . . is a special case of the social phenomenon of leadership" (1928, pp. 33–4).

innovations was at the heart of Schumpeter's business cycle theory.) As he stated: "The historic and irreversible change in the way of doing things we call 'innovation' and we define: innovations are changes in production functions which cannot be decomposed into infinitesimal steps. Add as many mail-coaches as you please, you will never get a railroad by so doing" (Schumpeter, 1935, p. 7).

Schumpeter's emphasis upon the centrality of creative destruction as an integral part of the capitalist growth process has been sharply criticized by Strassmann (1959a). Strassmann points out that in the period 1850 to 1914 at least, the old and new technologies coexisted peacefully, often for several decades. Indeed, he shows that for some of the most important innovations in power production, ferrous metallurgy, and other industries, output under the old technology continued to grow in absolute terms long after the introduction of the new technology. (See also Strassmann, 1959b.)

In sharp contrast to Schumpeter's emphasis upon the discontinuous nature of technical progress – a view that left a strong imprint on an entire generation of professional economists – is another school of thought that has been more impressed with continuity in technological change. Many aspects of this perspective may be traced back to Marx, who was, after all, a contemporary of Darwin, and who pointed acutely to the evolutionary elements in machine design. Marx also emphasized the larger social forces at work in technical progress and minimized the role of individuals. As he pointed out: "A critical history of technology would show how little any of the inventions of the eighteenth century are the work of a single individual" (Marx, 1867, p. 406). Marx's views on the nature of technical progress are examined in some detail in Rosenberg (1976).

The foremost and most carefully articulated expression, in the twentieth century, of the view of technical progress that emphasizes continuity appears in the work of A. P. Usher (1954; the first edition appeared in 1929). Usher called attention not only to the elements of continuity but also to the cumulative significance, in the inventive process, of large numbers of changes, each one of small magnitude. Moreover, and also in contrast to Schumpeter, who was primarily concerned with the consequences of inventions and not their origins, Usher was very much concerned with analyzing the nature of the inventive process and the forces that influenced events at the technical level. Usher's concern with the emergence of novelty in history led him to pay careful attention to the factors that conditioned or set the stage for a particular inventive breakthrough.

There has been an interesting attempt to merge and reconcile some of the useful elements in the works of Schumpeter and Usher. In undertaking this reconcilation, Vernon Ruttan also attempted to clarify the three related but distinct concepts of invention, innovation, and technological change (Ruttan, 1959). In doing so, he suggests how Usher's theory may be used to complement Schumpeter's where the latter's theory is weak and perhaps defective.

The view of technical progress as consisting of a steady accretion of innumerable minor improvements and modifications, with only very infrequent major innovations, was nicely embodied by S. C. Gilfillan in his book *Inventing the Ship* (1935a; see also his companion volume, 1935b). Although Gilfillan was primarily concerned with the social rather than the economic aspects of the process, his book provides a valuable close-up view of the gradual and piecemeal nature of technical progress, drawing heavily upon small refinements based upon experience and gradually incorporating a succession of improved components or materials developed in other industries. His analysis of the evolution of marine engines (chap. 2) is that of a slow sequence incorporating the growing strength and steam-raising capacity of boilers, the increasing reliance upon steel components as steel became cheaper, and the adoption of petroleum lubricants. In his discussion of the technological component of steamboat history in America, Louis Hunter in his book *Steamboats on the Western Rivers* (1949) also stresses the innumerable minor improvements and adaptations of an anonymous multitude of craftsmen, foremen, and mechanics.

Albert Fishlow's incisive study of productivity and technical progress in the American railroad system between 1870 and 1910 included an attempt to quantify the role of the separate factors at work in raising productivity and reducing costs (Fishlow, 1966). Productivity growth during this period was extremely high, and there were some important inventions, such as air brakes, automatic couplers and signaling devices, and the substitution of steel for iron rails. Nevertheless, Fishlow finds that the largest contribution to cost reduction by far was due to a succession of improvements in the design of locomotives and freight cars, even though the process included no readily distinguishable or memorable inventions. Nevertheless:

> Its cumulative character and the lack of a single impressive innovation should not obscure its rapidity. Within the space of some forty years – from 1870 to 1910 – freight car capacity more than trebled. The remarkable feature of the transition was its apparent small cost; capacity increased with only a very modest increase in dead weight, the ratio changing from 1:1 to 2:1. Over the same interval, locomotive force more than doubled. [Fishlow, 1966, p. 635]

Broadly similar findings have been reported by other scholars. In his study of the sources of increased efficiency in DuPont's rayon plants, Samuel Hollander concluded that the cumulative effects of minor technical changes upon cost reduction were greater than the effects of major technical changes (Hollander, 1965). Similarly, John Enos has studied the introduction of four major technical processes in petroleum refining in the twentieth century: thermal cracking, polymerization, catalytic cracking, and catalytic reforming (Enos, 1958; for a more detailed presentation, see Enos, 1962). Enos found that the cost reductions achieved by the later *improvements* in the major innovations were far greater than the cost reductions associated with their initial introduction. He concludes: "The evidence from the petroleum refining industry indicates that improving a process contributes even more to technological progress than does its initial development" (Enos, 1958, p. 180).

The rate of technical progress

One of the central historical questions concerning technical progress is its extreme variability over time and place. One of the most compelling facts of history is that there have been enormous differences in the capacity of different societies to generate technical innovations that are suitable to their economic needs. Moreover, there has also been extreme variability in the willingness and ease with which societies have adopted and utilized technological innovations developed elsewhere. And, in addition, individual societies have themselves changed markedly over the course of their own separate histories in the extent and intensity of their technological dynamism. Clearly, the reasons for these differences, which are not yet well understood, are tied in numerous complex and subtle ways to the functioning of the larger social systems, their institutions, values, and incentive structures. The explanation of these differences is intimately tied to such even larger questions as why social change occurs and why economic growth proceeds at such different speeds over time and place.

These questions have, of course, been addressed directly or indirectly by the major figures in social history and theory. To Karl Marx, technological dynamism was directly associated with the historical emergence of capitalist institutions. In Marx's view, capitalism leads to an immense expansion in productivity because the system creates uniquely powerful incentives and institutions for accelerating both technological change and capital accumulation. As Marx and Engels assert in *The Communist Manifesto,* the bourgeoisie "has been the first to show what man's activity can bring about. It has accomplished wonders far surpassing Egyptian pyramids, Roman aqueducts, and Gothic cathedrals" (Marx and Engels, 1848, vol. I, p. 35). The reason for this is that the capitalist class is the first ruling class in history whose interests are indissolubly linked to technological change instead of the maintenance of the status quo. As stated, again, in *The Communist Manifesto:* "The bourgeoisie cannot exist without constantly revolutionizing the instruments of production, and thereby the relations of production, and with them the whole relations of society. Conservation of the old modes of production in unaltered form was, on the contrary, the first condition of existence for all earlier industrial classes" (Marx and Engels, 1848, vol. I, p. 36). Yet, although Marx examines the historic rise of capitalism in response to expanding profit-making opportunities in the sixteenth century and scrutinizes capitalist institutions with extreme care, he does not really offer a satisfactory account of why capitalism emerged in Europe and not elsewhere. Although Marx argues persuasively that rapid technical progress in the West has, historically, been inseparable from capitalism, he does not really explain why this institutional vehicle for rapid technical progress did not emerge in such places as the Near East or Asia. This apparent lacuna in the Marxian analysis has been critically explored in Wittfogel (1957).

In accounts of the rise of capitalism in the West, a major theme has been the role of religion and its influence upon human behavior. Ever since Max Weber published his famous essay, *The Protestant Ethic and the Spirit of Capitalism,* in 1904–5, the nature of the association has been – and continues to be – hotly debated. In response to the view that Protestantism, with what Weber described as its "inner-worldly asceticism," promoted capitalism, it has been counterargued that capitalism can just as easily be regarded as the creator of Protestantism – insofar as Protestantism (especially its Calvinist variant) offered a highly congenial set of beliefs to the successful capitalist, who therefore embraced it with alacrity. Moreover, it has also been pointed out that capitalism had emerged in places with dominant Catholic populations, most notably in Italy and portions of Germany.

Recently, a new element has been introduced into this debate by Lynn White, who has contrasted Christianity as a whole to other religions (White, 1967). Seen in the broad context of the world's major religions, Christianity was unique in cultivating an activist and manipulative view toward the natural world.

> Especially in its western form, Christianity is the most anthropocentric religion the world has seen . . . Christianity, in absolute contrast to ancient paganism and Asia's religions (except, perhaps, Zoroastrianism), not only established a dualism of man and nature but also in-

> sisted that it is God's will that man exploit nature for his proper ends . . . By destroying pagan animism, Christianity made it possible to exploit nature in a mood of indifference to the feelings of natural objects. [White, 1967, p. 1205]

In this fashion, White simultaneously attributes the technological dynamism of postmedieval Europe and the recent ecological crisis to "the Christian axiom that nature has no reason for existence save to serve man" (p. 1207).

White's views on the critical role of Christianity in justifying an exploitive approach to the natural environment are spelled out in somewhat greater detail in his article "What Accelerated Technical Progress in the Western Middle Ages?" (White, 1963). He devotes more particular attention here to the labor-saving and power-exploiting aspects of Western European technology.

> The 19th century revulsion against abuses symbolized in Blake's "dark Satanic mills" has blinded historians to the fact that Western labour-saving power technology is profoundly humane in intent, and is largely rooted in religious attitudes. Its ideology is the Christian doctrine of man as developed not in the context of Greek contemplative intellectualism but rather in the framework of Latin voluntarism. The power machines of the Western Middle Ages which amazed Bessarion were produced in part by a spiritual repugnance towards subjecting anyone to drudgery which seems less than human in that it requires the exercise neither of intelligence nor of choice. The Western Middle Ages, believing that the Heavenly Jerusalem contains no temple (Rev., xx:22), began to explore the practical implications of this profoundly Christian paradox. Although to labour is to pray, the goal of labour is to end labour. [White, 1963, p. 291]

White points out further: "The cult of saints smashed animism and provided the cornerstone for the naturalistic (but not necessarily irreligious) view of the world which is essential to a highly developed technology" (p. 291).

White's book *Medieval Technology and Social Change* (1962) is, by all odds, the best single introduction to technological developments during the Middle Ages. The book deals with the origins of feudal institutions in terms of military technology (especially the role of the stirrup), the agricultural innovations associated with the northward shift of European civilization after A.D. 700 or so, and the growing reliance upon mechanical power and power-driven devices.

An excellent, concise statement of White's views at an earlier date, which shows more clearly his indebtedness to the seminal work of Lefebvre des Noettes on the exploitation of horsepower and to the great

French medievalist Marc Bloch, may be found in his article "Technology and Invention in the Middle Ages" (White, 1940). This article remains the best short introduction to the subject for the millennium of Western history between the collapse of the Roman Empire and the discovery of the New World.

White's highly provocative views have not gone unchallenged. See, for example, the article by Sawyer and Hilton (1963), which particularly questions the chronological basis of White's argument and challenges the deductions he has drawn to support his views from very limited evidence.

David Landes, in his authoritative book *The Unbound Prometheus* (1969), has attempted to reassess the reasons for European (especially British) technological dynamism. He asks what was unique in the pattern of European development that would explain why modern industrial technology emerged first in Western Europe. Landes identifies two distinctive characteristics. First, Europe experienced a pattern of political, institutional, and legal development that provided an especially effective basis for the operation of private economic enterprises. Systematic limitations were placed upon the arbitrary exactions of the state. Legal institutions afforded increasing protection and security to property. Contractual arrangements enforceable by the courts replaced the exercise of force or superior status. The uncertainties inherent in the unconstrained exercise of political power were progressively reduced. Although it may be argued that the difference was only one of degree, the emerging commercial classes in Western Europe were far freer from the arbitrary exercise of political power, their property was more secure against possible confiscation, and the business community was less inhibited by legal and other restrictions upon their freedom of action. Landes also argues that, with respect to many of the social variables that gave Europe an advantage over the rest of the world, England was in a superior position to the rest of Europe. The English class structure permitted a greater degree of mobility than existed in Europe, the English had greater success in limiting the power and privileges of the sovereign and nobility, and so on.

The second distinctive aspect of European development, according to Landes, was the high value placed upon the rational manipulation of the environment. European culture stressed the rational adaptation of means to ends. It created a culture in which superstition and magic were progressively deemphasized. This scientific revolution, as we now sometimes refer to it, was a uniquely European event.

Although it seems clear that, historically, modern science grew up and began to flourish alongside those major events that we label the

Renaissance, the Reformation, and the rise of capitalism, it can hardly be suggested that we understand the precise interconnections between these events – science, Renaissance, Reformation, capitalism – nearly as well as we should like. Moreover, it can hardly be argued that European civilization possessed a monopoly on rationality. Indeed, the monumental labors of Joseph Needham and his associates (Needham, 1954–) strongly suggest that Chinese civilization was technologically more advanced than that of Europe until perhaps the fifteenth century and was generally more successful in applying knowledge concerning natural phenomena to basic human needs. In fact, Needham goes even further in a highly suggestive article, "Science and Society in East and West" (Needham, 1969, Chap. 6). He states: "In many ways I should be prepared to say that the social and economic system of medieval China was much more rational than that of medieval Europe" (p. 197). Needham contrasts European feudalism, with its peculiar hereditary rules for selecting leaders, to the Chinese system, which based entry into the imperial bureaucracy upon a competitive examination system and was thus able to recruit talent from a far broader social base than was the case in Europe, with its hierarchies of enfeoffed barons. But the mandarin class, although it was indeed a form of meritocracy opposed to hereditary or aristocratic principles, was also hostile to wealth and acquisitive values. Chinese values, laws, and institutions remained dominated by scholar–bureaucrats in ways that provided neither the motivations nor the freedom of action that might give rise to a capitalist class with the capacity to transform society along lines required for the exploitation of new technologies.

It may be argued that what Europe possessed was a different kind of rationality: a readiness to learn and to borrow from other cultures, especially in matters technological. A. R. Hall has made the point forcefully:

> Perhaps European civilization could not have progressed so rapidly had it not possessed a remarkable faculty for assimilation – from Islam, from China, and from India. No other civilization seems to have been so widespread in its roots, so eclectic in its borrowings, so ready to embrace the exotic. Most have tended (like the Chinese) to be strongly xenophobic, and to have resisted confession of inferiority in any aspect, technological or otherwise. Europe would yield nothing of the pre-eminence of its religion and but little of its philosophy, but in processes of manufacture and in natural science it readily adopted whatever seemed useful and expedient. From the collapse of the Roman empire onwards there is indeed a continuous history of technological change in Europe, slight at first, but gradually becoming more

swift and profound. It would therefore be idle to discuss how this began, for it has always existed. [Hall, 1957, pp. 716–17]

A final important issue regarding the determinants of the rate of technical progress is the role of science. There is widespread agreement that the dependence of technical progress upon science has increased substantially over the past century or so. Aside from agreement on this trend, however, there is considerable controversy over the extent of that dependence in both the remote past and the modern period. The most forceful exponents of the view that technical progress was heavily dependent upon science at an early period are A. E. Musson and E. Robinson, in their book *Science and Technology in the Industrial Revolution* (1969). Musson and Robinson have brought together a wealth of evidence showing that in England, many intimate networks linked scientists with the business community. (See also A. E. Musson and E. Robinson, 1960, and A. E. Musson's extended introduction to *Science, Technology and Economic Growth in the 18th Century,* 1972.) On a more general plane, W. W. Rostow has emphasized the economic importance of science in postmedieval Europe in several books (Rostow 1952, 1960, 1975). On the other side are numerous historians of technology who have stressed the crude and patient empiricism, the trial-and-error approaches, and the ad hoc solutions of long generations of uneducated technologists, certainly before the mid-nineteenth century or so. These include Usher, Landes, and Gilfillan, whose works have been mentioned in other contexts. As A. R. Hall has stated about technical change in the century preceding 1760: "We have not much reason to believe that, in the early stages, at any rate, learning or literacy had anything to do with it; on the contrary, it seems likely that virtually all the techniques of civilisation up to a couple of hundred years ago were the work of men as uneducated as they were anonymous" (Hall, 1963). The relevant issues in this debate are examined with admirable clarity and judiciousness in Peter Mathias, "Who Unbound Prometheus? Science and Technical Change, 1600–1800" (1972). Clearly, the issues turn, in part, upon definitions and how rigorously one defines science. If one means systematized knowledge within a consistently integrated theoretical framework, the role of such knowledge is likely to have been small before the twentieth century. On the other hand, if one defines science more loosely in terms of procedures and attitudes, including the reliance upon experimental methods and an abiding respect for observed facts, it is likely to appear universal. What is certainly clear and is borne out by the histories of England, France, the United States, Japan, and Russia over the past two and a half centuries or so is that a top-quality scientific establishment and a high degree of scientific originality have been neither a necessary nor a sufficient condition for technological dynamism.

For eighteenth-century Britain, much attention has recently been devoted to the emergence of provincial scientific societies as institutional centers that created a close relationship between science and industry. Indeed, Robert Schofield has advanced the claim, based upon his careful study of the Lunar Society of Birmingham, that that society actually "represented an eighteenth-century technological research organization" (p. 415, 1957; see also Schofield, 1963). Charles Gillispie expresses his skepticism of the view "that theoretical science exerted a fructifying and even a causative influence in industralization" (1957, p. 399). Gillispie, who has focused upon the French experience, suggests that the interrelations between science and industry were far more complex than the view that explains industrial leadership in terms of earlier scientific attainments. He refers approvingly to L. J. Henderson's pithy observation that science has been far more indebted to the steam engine than the steam engine has been to science, and he advances the possibility that the interest of British scientists in industrial questions in the second half of the eighteenth century may have been more a *result* of British industrial leadership than a cause.

For an extremely useful study of the history of science, J. D. Bernal's four-volume *Science in History* (1971) should be consulted. Bernal deliberately seeks out, and is sensitive to, the interrelations between the scientific enterprise and larger social and economic forces. Whereas most historians of science are currently inclined to treat science as a largely autonomous phenomenon moving along in response to internal forces, Bernal, as a Marxist, looks upon scientific activity as something that is shaped by the larger society, as well as vice versa. Marx's own views on the history of science, and the ways in which scientific progress influences technical progress, are examined in Rosenberg (1974a).

The direction of technical progress

In addition to the rate of technical progress, there is the separate question of the direction of inventive activity. "Direction" here may take on a variety of dimensions, and we have already introduced the distinction between inventive activity that is directed toward product improvement or entails the invention of a new product, and inventive activity that is cost-reducing – or process invention. Many other distinctions are possible. A major concern among economists is the factor-saving bias of an invention. Many years ago, Hicks (1932) argued that inventions are "naturally" directed to reducing the utilization of a factor that is becoming relatively expensive. Thus, "The general tendency to a more rapid increase in capital than labour which has marked European history during the last few centuries has naturally provided a stimulus to labour-saving invention" (pp. 124–5).

More recently, the Hicksian view has been challenged by economists such as William Fellner, Paul Samuelson, and W. E. G. Salter on the grounds that rational businessmen always welcome cost reductions and that there is no reason why attention should focus upon invention possibilities with any particular factor-saving bias. As Salter put it:

> If . . . the theory implies that dearer labour stimulates the search for new knowledge aimed specifically at saving labour, then it is open to serious objections. The entrepreneur is interested in reducing costs in total, not particular costs such as labour costs or capital costs. When labour costs rise, any advance that reduces total cost is welcome, and whether this is achieved by saving labour or capital is irrelevant. [Salter, 1960, pp. 43–4]

This argument has challenged the widely accepted view of a labour-saving bias in Western – and especially North American – historical development. It seems reasonable to state that this disagreement between the theorists and the economic historians remains unresolved. The theorists insist that rational decision makers *under competitive conditions* (the qualification is critical, as Fellner recognized) would not deliberately engage in a biased search, whereas the economic historians have continued to assume and to suggest further historical evidence of an apparent labor-saving bias.

This confrontation became apparent after the publication of H. J. Habakkuk's seminal book, *American and British Technology in the 19th Century* (1962). Habakkuk's book, which is a sustained attempt to account for the divergent technological experiences of America and Britain, is perhaps the most influential book in economic history in the past twenty years. Although it is introduced here in a discussion of the possible factor-saving bias of technology, this is, inevitably, somewhat arbitrary. For in that book, several distinct issues are considered in overlapping ways: (1) the choice of an optimal technique from among a range of existing alternatives, (2) the existence of an economic mechanism that leads to innovations with specific factor-saving biases, and (3) the rate of inventive activity. In Habakkuk's formulation, there is an intimate link between the labor-saving bias that he attributes to American nineteenth-century technological progress and the fact that technological progress in America appeared to be far more rapid than in Britain. In Habakkuk's view, labor scarcity and resource abundance led to a search for labor-saving inventions – a search that was conducted at the capital-intensive end of the spectrum. Given the mechanical skills and the state of technical knowledge at the time, the possibilities for invention were richest at this end of the spectrum, and it was America's good fortune to have the resource endowment that pushed it forcefully in that direction. Thus, for Habakkuk, arguments concerning the rate and direction of inventive activity are necessarily intertwined.

The conflict between Habakkuk's argument, which is highly convoluted and has been drastically oversimplified above, and neoclassical economic reasoning is still unresolved. Peter Temin has twice attempted a rigorous reformulation of the issues – in "Labor Scarcity and the Problem of American Industrial Efficiency in the 1850s" (1966) and "Labor Scarcity in America" (1971). A short but useful survey of much of the "fallout" from Habakkuk's book is given in S. B. Saul's editor's introduction to *Technological Change: the U.S. and Britain in the 19th century* (1970). More recently, Paul David has attempted a major reformulation of the Habakkuk argument, attempting to reestablish the connection between America's resource abundance and the nature of the country's technical development (1975, chap. 1). In David's model, relative factor prices influence choices among techniques but such choices, in turn, have a strong influence upon the path of subsequent technological change. David employs here a localized learning-by-doing argument, one in which the decisions concerning techniques influence the later learning process. Thus, the choice decisions turn out to be far more fateful than is usually recognized because they set in motion a long-run evolutionary process linking factor prices, the choice of techniques, and the direction of technological change. It should be noted that, for David, an understanding of technological change becomes inseparable from its history. As he puts it:

> Because technological "learning" depends upon the accumulation of actual production experience, short-sighted choices about what to produce, and especially about how to produce it using presently known methods, also in effect govern what subsequently comes to be learned. Choices of technique become the link through which prevailing economic conditions may influence the future dimensions of technological knowledge. This is not the only link imaginable. But it may be far more important historically than the rational, forward-looking responses of optimizing inventors and innovators, which economists have been inclined to depict as responsible for the appearance of market- or demand-induced changes in the state of technology. [P. 4]

Some empirical support for the David model may be found in Rosenberg (1969a), especially the manner in which the problem-solving activities of technically trained personnel reinforce the localized learning hypothesis.

An important attempt to synthesize a wide variety of historical literature on the interaction between resources and agricultural technology is presented in the book *Agricultural Development,* by Vernon Ruttan and Yujiro Hayami (1971). The authors develop a theoretical framework for examining the patterns of agricultural development in individual countries within which endogenous technological change plays a critical role. Central to their approach is a theory of induced innovation that incorporates a unique, dynamic response of each country to its agricultural resources and input prices. Ruttan and Hayami argue that

> there are multiple paths of technological change in agriculture available to a society. The constraints imposed on agricultural development by an inelastic supply of land may be offset by advances in biological technology. The constraints imposed by an inelastic supply of labor may be offset by advances in mechanical technology. The ability of a country to achieve rapid growth in agricultural productivity and output seems to hinge on its ability to make an efficient choice among the alternative paths. Failure to choose a path which effectively loosens the constraints imposed by resource endowments can depress the whole process of agricultural and economic development. [Pp. 53–4]

In developing their induced innovation model, Ruttan and Hayami postulate the existence of a "metaproduction function." This is an envelope curve that goes beyond the production possibilities attainable with existing knowledge and described in a neoclassical long-run envelope curve. It describes, rather, a locus of production possibility points that can be *discovered* within the existing state of scientific knowledge. Points on this surface are attainable, but only at a cost in time and resources. They are not presently available in blueprint form.

Within this framework, Ruttan and Hayami study the growth of agricultural productivity among countries, with special emphasis upon the contrasting experiences of Japan and the United States, as an adaptive response to altering factor and product prices. The adaptation process is conceived as the ability to move to more efficient points on the metaproduction function, especially in response to the opportunities being generated by industry, which offer a potential flow of new inputs. The Ruttan and Hayami approach is the most detailed attempt to date to describe the mechanisms through which technological innovation and adaptation take place in response to shifting patterns of relative resource scarcities.

Our understanding of the factors determining the direction of technological change has been enhanced by the work of the late Jacob Schmookler. Whereas earlier work in economics tended to treat technological change as an exogenous variable, something that had important economic consequences but no readily identifiable economic causes,

Schmookler's work has served to demonstrate that the direction of technological change is responsive to economic forces – and that, indeed, technological change is an economic activity and can be usefully studied as such. In his book *Invention and Economic Growth* (1966), Schmookler marshaled a vast body of historical data to support his view that the allocation of resources to inventive activity is determined primarily by demand-side forces. In examining the American railroad industry, for which comprehensive data are available for over a century, Schmookler found a close correspondence between increased purchases of railroad equipment and components and slightly lagged increases in inventive activity as measured by new patents on such items. The lag is highly significant because, Schmookler argues, it indicates that variations in the sale of equipment induce variations in inventive effort. Schmookler finds similar relationships in building and petroleum refining, although the long-term data on these industries are less satisfactory. Moreover, in examining cross-sectional data for many industries in the years before and after the Second World War, Schmookler finds a very high correlation between capital goods invention for an industry and the sale of capital goods to that industry. These data support the view that for inventors, increased purchases of equipment by an industry signal the increased profitability of inventions in that industry, and direct their resources and talents accordingly. Thus, Schmookler concludes that demand, by influencing the size of the market for particular classes of inventions, is the decisive determinant of the allocation of inventive effort. Several of Schmookler's articles, as well as the patent data upon which his work was based, have been reprinted in Schmookler (1972). The data in this volume consist of more than 400 times series for patents granted, going as far back as 1837, and classified by type and industry of probable use.

In emphasizing the importance of demand-side forces in determining the direction of technological change, Schmookler sometimes neglected the importance of supply-side factors or made highly simplifying assumptions about their role. An article by Nathan Rosenberg (1974b) attempts to introduce some of the complexities of supply-side variables into Schmookler's framework and to suggest how these might influence the outcome of his arguments. Some closely related arguments are made in a short but wide-ranging paper by William Parker (1961). Parker looks at the timing and the sequence of technological changes over the past two centuries. In the process, he offers some provocative suggestions on how the state of fundamental science and differences in the complexity of classes of natural phenomena may have shaped the history of industrial change.

The diffusion of new technologies

For several decades, many historians, even economic historians, have focused their attention overwhelmingly upon one aspect of the question of technical progress: "Who did it first?" They have considered the claims to priority of different individuals. Such questions are, indeed, important to the history of *invention*. Much less attention, however, if any at all, has been accorded to the rate at which new technologies have been adopted and embedded in the productive process. Indeed, the diffusion process has often been assumed out of existence. This has been done by identifying the economic impact of an invention with the first date of its demonstrated technological feasibility or – what is hardly the same thing – the securing of a patent.

For the history of technical progress, however, these questions of priority are of secondary importance. Although the availability of a name and a date may simplify the writing of elementary histories, they add very little to our appreciation of the economic consequences of an invention. From the point of view of their economic impact, it is the diffusion process that is critical. That is because the productivity-increasing effects of superior technologies depend upon their utilization in the appropriate places. Happily, this point has been increasingly realized. As a result, the diffusion process is one of the most intensively explored subjects in economic history.

Much of the recent writing on the diffusion of inventions is the work of the so-called New Economic Historians, who have found out that when the necessary data are available, the diffusion process is admirably suited to quantitative analysis. Some work on diffusion of technology did, however, antedate the emergence of the New Economic History. There was Marc Bloch's admirable "Avenement et conquetes du moulin à eau" (1935). Bloch provided a masterly analysis, which turned primarily upon changing legal and economic conditions as they affected the availability of servile labor, of the lag of an entire millennium between the invention of the water mill and its widespread adoption. Several authors have noted the heavy dependence of technological diffusion in the past upon the geographic movement of skilled workers. See, for example, the illuminating account of a sixteenth-century transfer in Scoville (1951). Similarly, A. R. Hall has concluded: "It seems fairly clear that in most cases in the 16th century – and indeed long afterwards – the diffusion of technology was chiefly effected by persuading skilled workers to emigrate to regions where their skills were not yet plentiful" (Hall, 1967, p. 85). For a similar conclusion with respect to mining and metallurgy in an earlier period, see Gille (1963). Even when

Great Britain, "the Work-shop of the World," purchased a large quantity of American gun-making machinery in the 1850s for introduction into the government arsenal at Enfield, American machinists and supervisory personnel had to be employed (see Rosenberg, 1969b, Introduction). There is much evidence that the transmission of industrial technology from England to the Continent in the first half of the nineteenth century was also heavily dependent upon the same sort of personal mechanism. David Landes has suggested, for example, that in spite of legal prohibitions until 1825, there were at least 2,000 skilled British workers on the Continent providing indispensable assistance in adopting new techniques (Landes, 1969). Landes observes: "Perhaps the greatest contribution of these immigrants was not what they did but what they taught . . . The growing technological independence of the Continent resulted largely from man-to-man transmission of skills on the job" (p. 150). For a more detailed discussion of this subject, see Henderson (1954).

Recently, economic historians have begun to devote more attention to institutional factors as an influence on the pace of diffusion. This literature emphasizes the role of such factors as the lowering of transaction costs in improving the environment for innovation. This point is stressed in Davis and North (1971). The lowered cost of acquiring the necessary information about new technologies seems also to have been crucial in the timing of the diffusion of innovations (on this point, see Saxonhouse, 1974). Saxonhouse attributes the "superfast" diffusion of best-practice techniques in the Japanese textile industry to the low cost of acquiring information by textile firms. These low costs, in turn, he attributes to the actions of a business trade association, the use of a common capital goods supplier (Platt Brothers) with an energetic sales engineering staff, and a high degree of technical cooperation among firms. Additional studies of this sort should help greatly to improve our understanding of the institutional mechanisms involved in the diffusion of technology.

In a study of productivity change in ocean shipping before 1850, Douglass North finds that a superior cargo ship, the Dutch flute (or "flyboat"), which had been introduced into the Baltic route and the English coal trade in the first half of the seventeenth century, did not enter the Atlantic routes for over a century because of the prevalence of piracy (North, 1968). The economies resulting from modification of ship designs and the resulting reduction of crew size were not feasible so long as cargo ships had to carry armaments. The subsequent diffusion of new technologies in ocean shipping, especially the transition from sail to steam, may be traced through the following articles: North

(1958), Walton (1970–1), Graham (1956), Walton (1970), Harley (1971).

At the technological level, some attention has been given to the cumulative impact of numerous technical improvements, modifications, and adaptations in influencing the timing of the adoption of an innovation. The diffusion process is usually dependent upon a stream of improvements in the performance characteristics of an invention, its progressive modification and adaptation to suit the needs or specialized requirements of various submarkets, and upon the availability and introduction of other complementary inputs that make an original invention more useful. These and related points have been developed in Rosenberg (1972). The uniquely important historical role of the capital goods industries in facilitating this diffusion process (in addition to its central role in invention) has been explored in two articles by Nathan Rosenberg (1963a, 1963b).

Extensive econometric research on technological diffusion has been conducted in recent years, some of which goes back far enough to qualify as economic history. The seminal article was written by Zvi Griliches (1957), who has presented a less technical account of his findings in Griliches (1960). Griliches argues that the timing of the diffusion process can be explained very well in economic terms. He shows that the behavior of both farmers and hybrid-seed producers was firmly grounded in expectations of profit. The time lag in the first introduction of hybrid seed into a particular region, the rate at which farmers shifted to the new seed once it became available, and the extent to which the new seed replaced the old open-pollinated varieties all turned upon profitability. In areas where the profits were large and unambiguous, the transition was exceedingly rapid. In Iowa, where the hybrid corn was particularly well suited – and the profitability of the transition was accordingly high – it took farmers only four years to switch from 10 to 90 percent of their acreage to hybrid corn.

Edwin Mansfield has been the most prolific and influential contributor to the econometrics of the diffusion process. His work, like that of Griliches, is an attempt to identify and quantify the underlying economic variables that account for what we know of the diffusion process. An excellent introduction to this subject appears in Mansfield (1968, chap. 4). One of Mansfield's most important studies from the economic historian's perspective is "Technical Change and the Rate of Imitation" (1961). In that paper, Mansfield studies the speed with which twelve important innovations spread from firm to firm in four industries – bituminous coal, iron and steel, brewing, and railroads. His empirical results demonstrate the general overall slowness of the diffusion process and the wide variation in the speed with which particular techniques are adopted. His model also suggests, inter alia, that the rate of imitation was a direct function of the profitability of a given innovation and a decreasing function of the size of the investment required for its installation.

In an influential article, Paul David (1966) attempted to account for the fact that, although the reaper had been invented in the early 1830s, little diffusion occurred for many years. In the mid-1850s, however, fully two decades after its initial introduction, midwestern farmers discarded their old, labor-intensive techniques of cutting grain and adopted the reaper on a large scale. By using a threshold function relating to farm size, and by focusing upon the manner in which the rising relative cost of harvest labor lowered the threshold size, David accounts for both the earlier neglect and the rapid adoption of the reaper on family-sized farms during the 1850s. David's article has been criticized by Alan Olmstead (1975). Olmstead argues that the threshold model ignores the divisibility of machine services through sharing and contracting arrangements that, he asserts, were widely practiced. Olmstead argues that the reaper was adopted after twenty years because of the cumulative effect of numerous small improvements that gradually raised its productivity, making it commercially feasible in the 1850s.

Two excellent studies have attempted to account for the slow international diffusion of technologies by linking this diffusion to differences in environmental conditions. Such conditions include variations in the supplies of the factors of production but are not exhausted by them. The first study, again by Paul David, attempts to account for the forces impeding the mechanization of reaping in British agriculture in the second half of the nineteenth century (David, 1971). David argues that the state of the agricultural landscape in Britain presented significant barriers to the mechanization of field operations in the years after 1850. The barriers go far beyond the restrictive framework within which the "choice of technique" is typically discussed. British agriculture had to contend with topographical obstacles that were not encountered on the "broad, level and stone-free prairies of the Midwest." "Two features of the farming landscape in Britain must be considered as having obstructed the progress of mechanical reaping: the character of the terrain – which is to say, the nature of the field surfaces across which the implement would have to be drawn; and secondly, the size, shape and arrangement of the fields – or, more generally speaking, the layout of farms' cereal acreage" (David, 1971, pp. 148–9). These features include obstacles to the use of heavy, wheeled machinery presented by such vestiges of an earlier husbandry as drainage systems based upon ridge-

and-furrow contours, open irrigation trenches, small and irregularly shaped fields, separation of fields by hedges that hampered convenient movement of machinery, and so on. This situation was further complicated by legal and institutional arrangements that made farmers less willing and able to undertake necessary land improvements.

The second valuable study of international diffusion appears in the work of Peter Temin. Temin attempts to account for the long delay in the American adoption of coke in the blast furnace, a central innovation in the industrial revolution. Whereas over 90 percent of British blast furnaces were employing coke before 1810, Americans fifty years later were using it in just over 10 percent of their pig iron production. Temin's explanation is presented in terms of basic differences in resource endowments and the historical sequence in which the American resource base was progressively uncovered. To begin with, America had enormous amounts of wood, making it far less important than in Britain to substitute a mineral fuel for charcoal in the blast furnace. Moreover, although America possessed large quantities of high-quality coking coal, these were located west of the Appalachians, far from the main population centers in the antebellum period. Although there were substantial coal deposits in eastern Pennsylvania, they consisted of anthracite. The absence of gas in anthracite made ignition very difficult, so that anthracite could not be used with the blast furnace technology developed in eighteenth-century Britain. This situation began to change only in the 1830s with the development of Neilson's hot blast, which finally made it possible to introduce the readily accessible anthracite into the blast furnace. America's shift to a coke-smelting technology came only in the years after the Civil War and was associated with the westward movement of population (Temin, 1964a; see also Temin, 1964b).

Impact of technical progress upon productivity growth

In the past twenty years, economists and economic historians have attempted to develop serious quantitative measures of the contribution of technical progress to economic growth. This research is fraught with difficulties, both methodological and conceptual. Not only is it difficult to sort out the contribution of technical progress from other related contributions – capital formation, education, resource allocation – but there are no unambiguous measures of output over time periods long enough to permit large changes in both prices and the relative importance of each component of output (the index number problem). Nor are there satisfactory ways of taking account of significant quality changes or product innovations in the measures of an economy's changing output – and, as we have already seen, such changes are part and parcel of the impact of technical progress. Nevertheless, much time and effort have been devoted to these problems. Indeed, in recent years, the increased attention devoted by economic historians to technical progress has been due, in no small measure, to a recognition of the major role of technical progess in economic growth. This recognition may be dated to the papers by Moses Abramovitz (1956) and Robert Solow (1957). Both of these papers explored the quantitative importance of technical progress to the long-term economic growth of the American economy. They differed in several respects on such matters as time periods, coverage, and basic methodology. Nevertheless the authors concurred that only a very small portion of the long-term growth in American per capita output can be accounted for by an increasing quantity of capital and labor inputs, as these are conventionally measured. Both papers strongly suggested that the growth in per capita output has depended far more on increasing the productivity of resources than on using more resources. Although there was some tendency to label the increased productivity "technological change," such a label was not justified, as Abramovitz was careful to point out. This is because the extremely large residual with which both authors were left after attempting to measure the growth in output per capita that was attributable to rising inputs per capita encompassed a wide range of possible causes of improved efficiency other than technological change. In fact, the methodologies were such that the residual captured all causes of rising output per capita other than rising inputs per capita. The unexplained growth in resource productivity, as Abramovitz puts it, is a "measure of our ignorance," which turned out to be surprisingly large.

These startling results provoked a wide response, and numerous scholars have since tried their hand at explaining the components of the residual and the probable significance of each. Perhaps the most heroic and certainly the best-known attempt has been made by Edward Denison. Denison examined American economic growth between 1929 and 1957 and also between 1909 and 1929 and, with the use of certain simplifying assumptions, attempted to quantify the contribution of each of a number of variables to that growth. In addition to estimating the contributions of changing inputs of capital and labor, Denison attempted to adjust for quality changes in labor inputs, such as those attributable to more education and the effects of shorter work days on quality. Among Denison's most significant findings in estimating the components of the residual were the importance of advances in knowledge and the role of economies of scale (see Denison, 1962a, b.) Denison has applied essentially the same methodology to a massive research effort – a detailed comparative examination of the differing economic growth performances after the Second World War of the United States and eight European countries (Belgium, Denmark, France, Germany, Italy, the Netherlands, Norway, the United Kingdom) (see Denison, 1967).

These studies of technological change at a high level of aggregation have now been supplemented by more disaggregated ones. A pioneering effort to measure the social returns of a single innovation was the study of hybrid corn by Zvi Griliches. Griliches estimates that over the period 1910–55, the social rate of return on private and public resources committed to research on this highly successful innovation was at least 700 percent (Griliches, 1957). In a major attempt to sort out the determinants of productivity growth in American cereal production (wheat, corn, and oats), William Parker found that output per worker more than tripled between 1840 and 1911 (Parker and Klein, 1966). Parker concluded that 60 percent of this increase was attributable to mechanization, which raised the ratio of acreage to workers, and that practically all of the observed growth in productivity can be explained by the combination of mechanization and the westward expansion of agriculture. Parker also points out that the most important improvements came in those activities that had previously been highly labor-intensive – especially the harvesting and post-harvesting operations. In fact, two innovations alone – the reaper and the thresher – accounted for 70 percent of the total gain from mechanization. See also Parker's more wide-ranging treatment of the issues in Parker (1967).

A critical issue among New Economic Historians in examining industrial growth, productivity increases, and the role of technological change is to attempt to disentangle the roles of demand and supply factors in the process. Much earlier historical writing, for example, had simply taken it naively for granted that the observed rate of growth of an industry could be attributed completely to improvements in technical efficiency. But clearly, the rate of growth of an industry's output over time reflects the growth of demand- as well as supply-side factors, and econometric history attempts to provide quantitative statements of the respective importance of each. Robert Fogel and Stanley Engerman have developed a rigorous model that permits the identification and measurement of these separate influences in their article, "A Model for the Explanation of Industrial Expansion During the Nineteenth Century: With an Application to the American Iron Industry" (1969). With respect to the spectacular growth of the New England cotton textile industry after 1815, Robert Zevin found that supply-side factors accounted for less than half of the growth of output up to 1833 (Zevin, 1971). Zevin concludes that

> technical progress made a very modest contribution to the growth of total cotton cloth production from 1815 to 1833. Shifts in the demand curve alone would have caused production to expand at some 8 per cent or 9 per cent a year compared to a total growth rate of 15.4 per cent a year. This implies a supply induced growth of 6 per cent or 7 per cent a year. Approximately five sixths of this, in turn, can be attributed to technological developments. Hence during the eighteen years of the most revolutionary changes in the technology of cotton textile production, those changes, by themselves, could only have caused cloth production to expand at the fairly evolutionary pace of 5 per cent to 6 per cent a year.

Obviously, many more such studies are called for before we can claim to have a serious quantitative understanding of the impact of technological progress.

It is important to appreciate the role of interindustry relations in considering the contribution of technical progress to productivity growth. For example, the growth in agricultural productivity due to American westward expansion and the associated increasing degree of regional product specialization were, in turn, dependent upon technical improvements elsewhere – as in transport facilities. Indeed, by the end of the nineteenth century, the combined innovations of the railroad, the iron steamship, and refrigeration had created, for the first time in human history, a high degree of agricultural specialization on a worldwide scale. These developments are neatly spelled out in Youngson (1965).

The attempt to establish a close relationship between the historical time pattern of technical improvements in an innovation and the resulting productivity growth is full of pitfalls. Improvements will not necessarily be incorporated directly into the productive process, especially when they require the purchase of a new capital asset by the individual firm. Under such circumstances, as Salter has emphasized, the realization of technical improvements is tied to the age distribution of the existing capital stock and the rate at which new capital goods are being acquired (Salter, 1960).

A second reason for the lack of correspondence between technological improvement and actual productivity growth pertains to the period before which the technique has been widely adopted. When, in the early stages of its development, the cost of production with the new technique is very high, improvements leading even to significant cost reductions may have very little effect upon the rate of adoption. When, through accumulated improvements, the costs are eventually reduced and become roughly equivalent to those prevailing under the old technology, even a small further reduction may then lead to widespread adoption. Or, alternatively, at this point even relatively small changes in factor prices may shift the balance sharply in favor of this new technique, depending upon the nature of its factor-saving bias. That is to say, there is a threshold level at which the costs of the new technology become competitive with those of the old. (As we have already seen, the notion of a threshold has been exploited by Paul David with respect to farm size and the adoption of the reaper. David did not, however, emphasize the role of technological change in approaching and crossing the threshold in that particular instance. Rather, he stressed the rising cost of harvest labor relative to the price of reapers, which, in effect, lowered the threshold size of farms at which it became economic to assume the capital costs of the reaper.)

Thus, there may be a long gestation period in the development of a new technology during which gradual improvements are not exploited because the costs under the new technology are still substantially in excess of those of the old. However, as the threshold level is approached and eventually pierced, adoption rates of the new technology may become increasingly sensitive to further improvements. Thus, very large technological improvements may be made in an innovation during its "prenatal" period without any substantial repercussions. Conversely, even small further technological improvements made after the innovation has reached a threshold level may lead to rapid, large-scale productivity consequences.

Finally, we need to take account of other problems connected with measuring the productivity-increasing effects of technical progress – problems that have created a great deal of controversy in recent years. These effects will obviously depend, among other things, upon how widely an innovation is used throughout the economy – as indicated, for example, by the number of cells that it occupies in an input–output table. Of course it is true that, *ceteris paribus,* the more widely an innovation is used, the greater its aggregate productivity-increasing effects are likely to be. But that is only part of the story. An innovation may reduce costs only slightly below those of the old technology and may eventually become widely adopted. As a result, its ubiquitousness may superficially suggest something grossly misleading about the extent of its economic importance. Clearly, then, in order to assess the economic importance of an innovation, we need to analyze very carefully not only the number of cells of an input–output table that it occupies but also by how much the innovation reduces costs below those of the technology it replaced. To assess the economic significance of an innovation, we need to know not only that it reduced costs, which is fairly obvious, but the magnitude of the cost reduction.

This point has been central in the intense debate over the contribution made by the railroads to American economic growth in the nineteenth century. Whereas all participants in this debate have agreed upon the fundamental role that was played by the reduction in transportation costs, opinions have differed sharply over the role played by one particular innovation in transport – the railroad. It had previously been held that railroads were indispensable to American economic growth in the nineteenth century. It was believed not only that railroads were an efficient mechanism for providing transport services but that no effective alternatives existed or could have been produced. Robert Fogel has labeled this view "the axiom of indispensability" and has sharply criticized it.

> Evaluation of the axiom of indispensability . . . requires an examination of what the railroad did but also an examination of what substitutes for the railroad could have done. Railroads warrant the title of indispensability only if it can be shown that their incremental contribution over the next best alternative directly or indirectly accounted for a large part of the output of the American economy during the nineteenth century. Yet the historical evidence used to buttress the axiom is limited almost exclusively to descriptions of what the railroads did. There are few systematic inquiries into the ability of competing forms of transportation to have duplicated effects ascribed to railroads. As a consequence the range and potentiality of the supply of alternative opportunities are virtually unknown. The conclusion that the developmental potential of substitutes for railroads was very low therefore rests on a series of questionable assumptions rather than on demonstrated facts. [Fogel, 1964, p. 10]

Fogel's attack upon the axiom of indispensability essentially reduces to the proposition that railroads were not wildly superior to canals, an inference that has often been implicitly drawn from the evidence of the widespread adoption of the railroad. He undertakes with great energy and ingenuity to demonstrate what the American economy might have looked like by 1890 if the railroads had never come into existence. Fogel's calculations suggest that, had the railroads never existed, the gross national product of the United States would have been less than 5 percent below the level that it in fact attained that year. Fogel therefore does not deny that rapid economic development in the nineteenth century was heavily dependent upon reductions in transport costs. He denies that it was heavily dependent upon the railroad.

The essential conclusion is that the economic impact of an innovation must be examined in terms of the size of the cost reduction that it makes possible, and that this cost reduction can be assessed only by comparing the new technology with the cost structure of the alternative available technologies. Observations on the widespread use of the innovation, by themselves, are not enough.

Fogel's book touched off a heated debate. The most illuminating point of entry into this literature is Paul David's article, "Transport Innovation and Economic Growth: Professor Fogel on and off the Rails" (1969). For a study of England and Wales that attributes a considerably higher social saving to the railroad than Fogel found for the United States, see Hawke (1970).

It is ironic, in view of the storm of controversy touched off by Fogel's downward assessment of the role of the railroads, that his central idea was a quite uncontroversial one, namely, that *no* single technological innovation was essential to economic growth.

> The most important implication of this study is that no single innovation was vital for economic growth during the nineteenth century. Certainly if any innovation had title to such distinction it was the railroad. Yet, despite its dramatically rapid and massive growth over a period of a half century, despite its eventual ubiquity in inland transportation, despite its devouring appetite for capital, despite its power to determine the outcome of commercial (and sometimes political) competition, the railroad did not make an overwhelming contribution to the production potential of the economy. [Fogel, 1964, pp. 234–5]

Perhaps it may be appropriate to conclude this survey of the historiography of technical progress with a slight variant of the point that Fogel is making here. In a society that is capable of generating rapid technical progress, no single innovation is indispensable. However, the reason for this is not that individual innovations do not matter, in some absolute sense, but rather that such a society can readily generate *substitute* innovations. It is precisely the capacity to generate many possible innovations that renders any single innovation expendable. If this is correct, it makes even more intriguing the question, discussed in the earlier section on "The Rate of Technical Progress," of the social determinants of a society's capacity for generating technical progress in the first place. For on this most fundamental issue, our understanding remains, at best, rudimentary.

References

Abramovitz, M. (1956). "Resource and Output Trends in the United States since 1870." *American Economic Review Papers and Proceedings*, May.

Bernal, J. (1971). *Science in History*. MIT Press, Cambridge, Mass. 4 vols.

Bloch, M. (1935). "Avenement et conquetes du moulin à eau." *Annales d'histoire economique et sociale,* vii.

David, P. (1966). "The Mechanization of Reaping in the Ante-Bellum Midwest." In H. Rosovsky, ed., *Industrialization in Two Systems: Essays in Honor of Alexander Gerschenkron.* Wiley, New York.

(1969). "Transport Innovation and Economic Growth: Professor Fogel on and off the Rails." *Economic History Review,* December.

(1971). "The Landscape and the Machine: Technical Interrelatedness, Land Tenure and the Mechanization of the Corn Harvest in Victorian Britain." In D. McCloskey, ed., *Essays on a Mature Economy: Britain after 1840.* Methuen, London.

(1975). *Technical Choice, Innovation and Economic Growth.* Cambridge University Press, Cambridge.

Davis, L., and North, D. (1971). *Institutional Change and American Economic Growth.* Cambridge University Press, Cambridge.

Denison, E. (1962a). "United States Economic Growth." *Journal of Business,* August.

(1962b). *The Sources of Economic Growth in the U.S. and the Alternatives before Us.* Committee for Economic Development, New York.

(1967). *Why Growth Rates Differ.* The Brookings Institution, Washington, D.C.

Enos, J. (1958). "A Measure of the Rate of Technological Progress in the Petroleum Refining Industry." *Journal of Industrial Economics,* June.

(1962). *Petroleum, Progress and Profits.* MIT Press, Cambridge, Mass.

Fishlow, A. (1966). "Productivity and Technological Change in the Railroad Sector, 1840–1910." In *Output, Employment and Productivity in the U.S. after 1800,* Studies in Income and Wealth No. 30. National Bureau of Economic Research, New York.

Fogel, R. (1964). *Railroads and American Economic Growth,* Johns Hopkins University Press, Baltimore.

Fogel, R., and Engerman, S. (1969). "A Model for the Explanation of Industrial Expansion during the Nineteenth Century: With an Application to the American Iron Industry." *Journal of Political Economy,* May–June.

Gille, B. (1963). "Technological Developments in Europe: 1100 to 1400." In G. Metraux and F. Crouzet, eds., in *The Evolution of Science.* New American Library, New York.

Gillispie, C. (1957). "The Natural History of Industry." *Isis,* 48.

Gilfillan, S. (1935a). *Inventing the Ship.* Follett, Chicago.

(1935b). *The Sociology of Invention.* Follett, Chicago.

Graham, C. (1956). "The Ascendency of the Sailing Ship, 1850–1885." *Economic History Review,* August.

Griliches, Z. (1957). "Hybrid Corn: An Exploration in the Economics of Technological Change." *Econometrica,* October.

(1958). "Research Costs and Social Returns: Hybrid Corn and Related Innovations." *Journal of Political Economy,* October.

(1960). "Hybrid Corn and the Economics of Innovation." *Science,* 29 July.

Habakkuk, H. (1962). *American and British Technology in the 19th Century.* Cambridge University Press, Cambridge.

Hall, A. (1957). "Epilogue: The Rise of the West," in C. Singer et al., *A History of Technology,* vol. III. Oxford University Press, Oxford.

(1963). *The Historical Relations of Science and Technology,* inaugural lecture. London.

(1967). "Early Modern Technology to 1600." In M. Kranzberg and C. Pursell, eds., *Technology in Western Civilization,* vol. I. Oxford University Press, Oxford.

Harley, C. (1971). "The Shift from Sailing Ships to Steamships, 1850–1890: A Study in Technological Change and Its Diffusion." In D. McCloskey, ed., *Essays on a Mature Economy: Britain after 1840*. Methuen, London.

Hawke, G. (1970). *Railroads and Economic Growth in England and Wales 1840–1870*. Oxford University Press, Oxford.

Henderson, W. (1954). *Britain and Industrial Europe, 1750–1870*. Liverpool University Press, Liverpool.

Hicks, J. (1932). *Theory of Wages*. Macmillan, London.

Hollander, S. (1965). *The Sources of Increased Efficiency: The Study of Du Pont Rayon Plants*. MIT Press, Cambridge, Mass.

Hunter, L. (1949). *Steamboats on the Western Rivers*. Harvard University Press, Cambridge, Mass.

Kuznets, S. (1930). *Secular Movements in Production and Prices*. Houghton Mifflin, Boston.

(1953). *Economic Change*. Norton, New York.

(1971). *Economic Growth of Nations*. Belknap Press, Cambridge, Mass.

(1972). "Innovations and Adjustments in Economic Growth." *Swedish Journal of Economics*, 74.

Landes, D. (1969). *The Unbound Prometheus*. Cambridge University Press, Cambridge.

Mansfield, E. (1861). "Technical Change and the Rate of Imitation." *Econometrica*, October.

(1968). *The Economics of Technological Change*. Norton, New York.

Marx, K. (1867). *Capital*. Modern Library, New York, no date.

Marx, K., and Engels, F. (1848). *The Communist Manifesto*. As reprinted in K. Marx and F. Engels, *Selected Works*. Foreign Languages Publishing House, Moscow, 1951. 2 vols.

Mathias, P. (1972). "Who Unbound Prometheus? Science and Technical Change, 1600–1800." In P. Mathias, ed., *Science and Society, 1600–1900*. Cambridge University Press, Cambridge.

Musson, A, ed. (1972). *Science, Technology and Economic Growth in the 18th Century*. Methuen, London.

Musson, A., and Robinson, E. (1960). "Science and Industry in the Late 18th Century." *Economic History Review*, December.

(1969). *Science and Technology in the Industrial Revolution*. Manchester University Press, Manchester.

Needham, J. (1954–). *Science and Civilization in China*. Cambridge University Press, Cambridge. 7 vols. in 12 parts.

(1969). *The Grand Titration*. Allen & Unwin, London.

North, D. (1958). "Ocean Freight Rates and Economic Development, 1750–1913." *Journal of Economic History*, December.

(1968). "Sources of Productivity Change in Ocean Shipping, 1600–1850." *Journal of Political Economy*, September–October.

Olmstead, A. (1975). "The Mechanization of Reaping and Mowing in American Agriculture, 1833–1870." *Journal of Economic History*, June.

Parker, W. (1961). "Economic Development in Historical Perspective." *Economic Development and Cultural Change*, October.

(1967). "Sources of Agricultural Productivity in the 19th Century." *Journal of Farm Economics*, December.

Parker, W., and Klein, J. (1966). "Productivity Growth in Grain Production in the United States, 1840–60 and 1900–10." In Dorothy Brady, ed., *Output, Employment and*

Productivity in the United States after 1800. National Bureau of Economic Research, New York.
Rosenberg, N. (1963a). "Technological Change in the Machine Tool Industry, 1840–1910." *Journal of Economic History*, December.
(1963b). "Capital Goods, Technology and Economic Growth." *Oxford Economic Papers*, vol. 15.
(1969a). "The Direction of Technological Change." *Economic Development and Cultural Change*, October.
Ed. (1969b). *The American System of Manufactures*. Edinburgh University Press, Edinburgh.
(1972). "Factors Affecting the Diffusion of Technology." *Explorations in Economic History*, Fall.
(1974a). "Karl Marx on the Economic Role of Science." *Journal of Political Economy*, July–August.
(1974b). "Science, Invention, and Economic Growth." *Economic Journal*, September.
(1976). "Marx as a Student of Technology." *Monthly Review*, July–August.
Rostow, W. (1952). *The Process of Economic Growth*. Oxford University Press, Oxford.
(1960). *The Stages of Economic Growth*. Cambridge University Press, Cambridge.
(1975). *How It All Began*. McGraw-Hill, New York.
Ruttan, V. (1959). "Usher and Schumpeter on Invention, Innovation and Technological Change." *Quarterly Journal of Economics*, November.
Ruttan, V., and Hayami, Y. (1971). *Agricultural Development*. Johns Hopkins University Press, Baltimore.
Salter, W. (1960). *Productivity and Technical Change*. Cambridge University Press, Cambridge.
Saul, S., ed. (1970). *Technological Change: The U.S. and Britain in the 19th Century*. Methuen, London.
Sawyer, P., and Hilton, R. (1963). "Technological Determinism: The Stirrup and the Plough." *Past and Present*, April.
Saxonhouse, G. (1974). "A Tale of Japanese Technological Diffusion in the Meiji Period." *Journal of Economic History*, March.
Schmookler, J. (1966). *Invention and Economic Growth*. Harvard University Press, Cambridge, Mass.
(1972). *Patents, Invention, and Economic Change. Data and Selected Essays*, ed. Z. Griliches and L. Hurwicz. Harvard University Press, Cambridge, Mass.
Schofield, R. (1957). "The Industrial Orientation of Science in the Lunar Society of Birmingham." *Isis*, vol. 48.
(1963). *The Lunar Society of Birmingham*. Oxford University Press, Oxford.
Schumpeter, J. (1928). "The Instability of Capitalism." *Economic Journal*, September. As reprinted in N. Rosenberg, ed., *The Economics of Technological Change*. Penguin Books, Harmondsworth, 1971.
(1935). "The Analysis of Economic Change." *The Review of Economic Statistics*, May. As reprinted in *Readings in Business Cycle Theory*, Blakiston, Philadelphia, 1944.
(1939). *Business Cycles*. McGraw-Hill, New York. 2 vols.
(1942). *Capitalism, Socialism, and Democracy*. Harper & Row, New York.
Scoville, W. (1951). "Minority Migrations and the Diffusion of Technology." *Journal of Economic History*, Fall.
Solow, R. (1957). "Technical Change and the Aggregate Production Function." *Review of Economics and Statistics*, August.

Strassmann, W. (1959a). "Creative Destruction and Partial Obsolescence in American Economic Development." *Journal of Economic History.* September.
(1959b). *Risk and Technological Innovation.* Cornell University Press, Ithaca, N.Y.
Temin. P. (1964a). *Iron and Steel in Nineteenth-Century America.* MIT Press, Cambridge, Mass.
(1964b). "A New Look at Hunter's Hypothesis about the Ante-Bellum Iron Industry." *American Economic Review Papers and Proceedings,* May.
(1966). "Labor Scarcity and the Problem of American Industrial Efficiency in the 1850s." *Journal of Economic History,* September.
(1971). "Labor Scarcity in America." *Journal of Interdisciplinary History,* Winter.
Usher, A. (1954). *A History of Mechanical Inventions,* 2nd. ed. Harvard University Press, Cambridge, Mass.
Walton, G. (1970). "Productivity Change in Ocean Shipping after 1870: A Comment." *Journal of Economic History,* June.
(1970–1). "Obstacles to Technical Diffusion in Ocean Shipping." *Explorations in Economic History,* Winter.
White, L. (1940). "Technology and Invention in the Middle Ages." *Speculum, April.*
(1962). *Medieval Technology and Social Change.* Oxford University Press, Oxford.
(1963). "What Accelerated Technical Progress in the Western Middle Ages?". In A. Crombie, ed., *Scientific Change.* Basic Books, New York.
(1967). "The Historical Roots of our Ecologic Crisis." *Science,* 10 March.
Wittfogel, K. (1957). *Oriental Despotism.* Yale University Press, New Haven, Conn.
Youngson, A. (1965). "The Opening of New Territories." In H. Habakkuk and M. Postan, eds., *The Cambridge Economic History of Europe,* vol. VI: *The Industrial Revolution and After.* Cambridge University Press, Cambridge.
Zevin, R. (1971). "The Growth of Cotton Textile Production After 1815." In Robert Fogel and Stanley Engerman, eds., *The Reinterpretation of American Economic History,* Harper & Row, New York.

Marx as a student of technology

I

This paper will attempt to demonstrate that a major reason for the fruitfulness of Marx's framework for the analysis of social change was that Marx was, himself, a careful student of technology. By this I mean not only that he was fully aware of, and insisted upon, the historical importance and the social consequences of technology. That much is obvious. Marx additionally devoted much time and effort to explicating the distinctive characteristics of technologies, and to attempting to unravel and examine the inner logic of individual technologies. He insisted that technologies constitute an interesting subject, not only to technologists but to students of society and social pathology as well, and he was very explicit in the introduction of technological variables into his arguments.

I will argue that, quite independently of whether Marx was right or wrong in his characterization of the future course of technological change and its social and economic ramifications, his formulation of the problem still deserves to be a starting point for any serious investigation of technology and its ramifications. Indeed, the following statement by Marx, amazingly fresh over a century later, reads like a prolegomenon to a history of technology that still remains to be written:

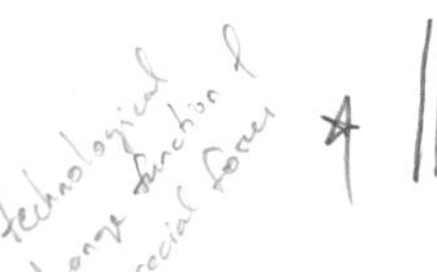

> A critical history of technology would show how little any of the inventions of the eighteenth century are the work of a single individual. Hitherto there is no such book. Darwin has interested us in the history of Nature's Technology, i.e., in the formation of the organs of plants and animals, which organs serve as instruments of production

This paper was originally published in *Monthly Review*, 28, July-August 1976, pp. 56–77.

In preparing this paper, I have had the considerable benefit of comments and suggestions from Jens Christiansen, Paul David, David Mowery, and Stanley Engerman. I am grateful also to the National Science Foundation for financial support.

> for sustaining life. Does not the history of the productive organs of man, of organs that are the material basis of all social organization, deserve equal attention? And would not such a history be easier to compile since, as Vico says, human history differs from natural history in this, that we have made the former, but not the latter? Technology discloses man's mode of dealing with Nature, the process of production by which he sustains his life, and thereby also lays bare the mode of formation of his social relations, and of the mental conceptions that flow from them.[1]

In what follows, I will focus first upon Marx's alleged technological determinism, second on his views on the characteristics of Modern Industry that are responsible for its high degree of technological dynamism, and finally on the special importance that Marx attaches to the role of the capital-goods sector in the generation of technological change. An important question regarding Marx's treatment of technology is, quite simply, what it was about his method or approach that made him so much more perceptive on this subject than any of his contemporaries. I think a few tentative methodological observations, upon which I will elaborate later, may be useful at the outset.

The method of historical materialism that Marx utilized was one that emphasized the interactions and conflicts of social classes and institutions, not individuals. Thus for Marx, invention and innovation, no less than other socioeconomic activities, were best analyzed as social processes rather than as inspired flashes of individual genius. The focus of Marx's discussion of technological change is thus not upon individuals, however heroic, but upon a collective, social process in which the institutional and economic environments play major roles.

Marx's historical approach was one that emphasized the discontinuous nature of social evolution, an evolutionary process that was, for him, moving forward under capitalism, just as it had under earlier forms of social organization. Rather than viewing capitalism as the final, logical outcome of a smooth, lengthy evolutionary process, Marx treated it as simply one stage in the process of historical evolution, while looking for the unique features of the capitalist forces of production and attempting to understand them in dynamic terms.

I would argue that the dialectical method is the most important factor in understanding the methodological basis for Marx's unique insights. Rather than positing some unidirectional chain of causation for technological change, Marx offers a far richer mode of analysis, one that

[1] Karl Marx, *Capital*, Vol. 1, Modern Library edition, p. 406, n. 2. All citations to the first volume of Capital are to this edition. The pagination is the same as in the Kerr edition.

emphasizes the mutual interactions and feedbacks between economy and technology. His analysis of the rise of the system of "machinofacture" and its implications for technological change, which I will examine in this paper, is an important example of the insights yielded by such a method.

II

First of all, it is necessary to advert briefly to the oft-repeated view that Marx was a technological determinist.[2] If by this we mean that technological forces are the decisive factor in generating socioeconomic changes – that technological factors are, so to speak, the independent variable in generating social change, which constitutes the dependent variable – it is easy to demonstrate that Marx subscribed to no such simplistic view. No doubt certain passages can be cited to support such an interpretation – most notably, of course, the statement from *The Poverty of Philosophy* that "The handmill gives you society with the feudal lord; the steam-mill, society with the industrial capitalist."[3] But such an interpretation, relying upon a few such aphoristic assertions, often tossed out in the heat of debate (as the quotation from *The Poverty of Philosophy* was thrown out in criticizing Proudhon), finds little support in Marx's own treatment of the major historical episodes with which he was concerned. Indeed, in a process no less central to Marx than the historical rise of capitalism itself, technological factors play no immediate role at all. For Marx, capitalism developed in Western Europe basically in response to growing markets and related opportunities for profit-making associated with the geographic explorations of the fifteenth century. The *locus classicus* for this view is, of course, the opening pages of the *Communist Manifesto:*

> The discovery of America, the rounding of the Cape, opened up fresh ground for the rising bourgeoisie. The East-Indian and Chinese mar-

[2] This view goes back more than fifty years in the professional economics literature. See Alvin Hansen, "The Technological Interpretation of History," *Quarterly Journal of Economics,* November 1921, pp. 72–83.

[3] Karl Marx, *The Poverty of Philosophy* (Moscow, n.d.), p. 105. The sentences that precede the one quoted above make Marx's meaning perfectly clear and reasonable. "Social relations are closely bound up with productive forces. In acquiring new productive forces men change their mode of production; and in changing their mode of production, in changing the way of earning their living, they change all their social relations." Moreover, as Marx points out later, "The handmill presupposes a different division of labor from the steam-mill" (p. 127). Surely one need not be a technological determinist to subscribe to these observations.

kets, the colonization of America, trade with the colonies, the increase in the means of exchange and in commodities generally, gave to commerce, to navigation, to industry, an impulse never before known, and thereby, to the revolutionary element in the tottering feudal society, a rapid development.

The feudal system of industry, under which industrial production was monopolized by closed guilds, now no longer sufficed for the growing wants of the new markets. The manufacturing system took its place. The guild-masters were pushed on one side by the manufacturing middle class; division of labor between the different corporate guilds vanished in the face of division of labor in each single workshop.

Meantime the markets kept ever growing, the demand ever rising. Even manufacture no longer sufficed. Thereupon, steam and machinery revolutionized industrial production. The place of manufacture was taken by the giant, Modern Industry, the place of the industrial middle class, by industrial millionaires, the leaders of whole industrial armies, the modern bourgeois.

Modern industry has established the world-market, for which the discovery of America paved the way. This market has given an immense development to commerce, to navigation, to communication by land. This development has, in its turn, reacted on the extension of industry; and in proportion as industry, commerce, navigation, railways extended, in the same proportion the bourgeoisie developed, increased its capital, and pushed into the background every class handed down from the Middle Ages.[4]

I have taken the liberty of quoting at length from a familiar source because this passage is, it seems to me, a definitive refutation of the view that Marx was a technological determinist. The question is: What are the factors that *initiate* change? What are the factors that cause other factors to change? A resolute technological determinist would presumably argue that the entire process of European expansion was initiated by navigational improvements that in turn generated the growth of overseas markets. But this is clearly not Marx's own view here. In fact, Marx states *twice* that the improvements in navigation were caused by the prior growth in markets and commercial opportunities. The passage makes it unmistakably clear that the technological changes associated with the two stages of capitalist development – the

[4] Karl Marx and Friedrich Engels, *The Manifesto of the Communist Party,* in Karl Marx and Friedrich Engels, *Selected Works,* Vol. 1 (Moscow, 1951), p. 34. See also *Capital,* p. 823; *The Poverty of Philosophy,* pp. 129–33; and Karl Marx, *Grundrisse* (New York, 1973), pp. 505–11.

manufacturing system and Modern Industry – were responses to an expanding universe of profit-making opportunities.

For Marx, then, capitalist relationships emerged when the growth of profit-making opportunities led to an expansion in the size of the productive unit beyond that which was characteristic of the medieval craft workshop. The mere quantitative expansion of such workshops led eventually to qualitative changes of a most basic sort in social relationships.[5] Although the system of manufacture totally dominated the first two and a half centuries of Western capitalism and led to major transformations in social relationships,[6] it was not associated with any major technological innovations. "With regard to the mode of production itself, manufacture, in its strict meaning, is hardly to be distinguished, in its earliest stages, from the handicraft trades of the guilds, otherwise than by the greater number of workmen simultaneously employed by one and the same individual capital. The workshop of the mediaeval master handicraftsman is simply enlarged."[7]

To regard Marx as a technological determinist, then, is tantamount to ignoring his dialectical analysis of the nature of historical change.[8] The essence of this view is that the class struggle, the basic moving force of history, is itself the product of fundamental contradictions between the forces of production and the relations of production. At any point in historical time, new productive forces emerge, not exogenously or as some mysterious *deus ex machina,* but rather as a dialectical outcome of a larger historical process in which *both* the earlier forces and relations of production play essential roles. As Marx forcefully put it: "It

[5] Marx, *Capital,* pp. 337–8, 367.

[6] "While simple co-operation leaves the mode of working by the individual for the most part unchanged, manufacture thoroughly revolutionizes it, and seizes labor power by its very roots. It converts the laborer into a crippled monstrosity, by forcing his detail dexterity at the expense of a world of productive capabilities and instincts; just as in the States of La Plata they butcher a whole beast for the sake of his hide or his tallow" (ibid., p. 396).

[7] Ibid., p. 353. Note also that Marx's panegyrics on the technological dynamism of capitalism apply, not to capitalism throughout its history, but only to capitalism as it existed in the century or so before the writing of the *Communist Manifesto.* "The bourgeoisie, *during its rule of scarce one hundred years,* has created more massive and more colossal productive forces than have all preceding generations together" (Marx and Engels, *Manifesto,* p. 37, emphasis added).

[8] For a forthright statement of technological determinism, see the work of the anthropologist Leslie A. White. According to White, a social system is "a function of a technological system." Furthermore, "Technology is the independent variable, the social system the dependent variable. Social systems are therefore determined by systems of technology; as the latter change, so do the former" (*The Science of Culture* [New York, 1971], p. 365).

must be kept in mind that the new forces of production and relations of production do not develop out of *nothing*, nor drop from the sky, nor from the womb of the self-positing Idea; but from within and in antithesis to the existing development of production and the inherited, traditional relations of property."[9]

Thus, for Marx the basic rhythm of human history is the outcome of this dialectical interaction between the forces and the relations of production. To categorize Marx as a technological determinist, one would have to demonstrate first that he does *not* intend his historical argument to proceed in a dialectical form. I think it is easy to demonstrate that he does.

III

If Marx was not a technological determinist he did, nonetheless, attach great importance to technological factors. The reasons for this are made clear in Chapter 7 of the first volume of *Capital*. Technology is what mediates between man and his relationship with the external, material world. But in acting upon that material world, man not only transforms it for his own useful purposes (that is to say, "Nature becomes one of the organs of his activity"[10]) but he also, unavoidably, engages in an act of self-transformation and self-realization. "By thus acting on the external world and changing it, he at the same time changes his own nature."[11] Technology, therefore, is at the center of those activities that are distinctively human. For technology comprises those instruments that determine the effectiveness of man's pursuit of goals that are shaped not only by his basic instinctive needs, but also those formulated and shaped in his own brain. "A spider conducts operations that resemble those of a weaver, and a bee puts to shame many an architect in the construction of her cells. But what distinguishes the worst architect from the best of bees is this, that the architect raises his structure in imagination before he erects it in reality. At the end of every labor

[9] Marx, *Grundrisse*, p. 278, emphasis Marx's. Elsewhere he states: "Whenever a certain stage of maturity has been reached, the specific historical form is discarded and makes way for a higher one. The moment of arrival of such a crisis is disclosed by the depth and breadth attained by the contradictions and antagonisms between the distribution relations, and thus the specific historical form of their corresponding production relations, on the one hand, and the productive forces, the production powers and the development of their agencies, on the other hand. A conflict then ensues between the material development of production and its social form" (Karl Marx, *Capital*, Vol. 3 [Moscow, 1959], p. 861).

[10] Marx, *Capital*, Vol. 1, p. 199. [11] Ibid, p. 198.

process, we get a result that already existed in the imagination of the laborer at the commencement."[12]

The informed student of society, therefore, can infer much concerning the nature of a society, its intellectual attainments, its organization, and its dominant social relationships by studying the instruments of human labor. But again it needs to be insisted that Marx's position here cannot be reduced to a crude technological determinism. In a highly perceptive passage that is sometimes cited as evidence of his technological determinism, Marx is, in fact, pointing to what can be *inferred* about the nature of earlier societies from their remaining artifacts. "Relics of by-gone instruments of labor possess the same importance for the investigation of extinct economical forms of society, as do fossil bones for the determination of extinct species of animals. It is not the articles made, but how they are made, and by what instruments, that enables us to distinguish different economical epochs. Instruments of labor not only supply a *standard* of the degree of development to which human labor has attained, but they are also *indicators* of the social conditions under which that labor is carried on."[13] I believe this passage is evidence of technological determinism in about the same sense that one is entitled to say that a thermometer *determines* body temperature or a barometer *determines* atmospheric pressure. All such statements are equally specious and misleading in mixing up measurement with causation.

The decisive technological changes with which Marx is concerned began around the middle of the eighteenth century. It is at this point that Britain began her transition from an industrial system of manufacture to what Marx calls Modern Industry. The general outline of his analysis, as presented in Chapters 13–15 of *Capital,* is well known and need not be repeated here. Rather, I will focus on certain specific features that, I believe, are still insufficiently appreciated.

First of all, Marx posed and dealt with a basic question concerning the nature of technology that has never received the attention it deserves. It is widely accepted that modern capitalist societies have achieved high levels of productivity because of the systematic application of scientific

[12] Ibid.

[13] Ibid., p. 200, emphasis added. The paleontological mode of reasoning is continued after the passage quoted. Marx adds in a footnote on the same page: "However little our written histories up to this time notice the development of material production, which is the basis of all social life, and therefore of all real history, yet prehistoric times have been classified in accordance with the results, not of so-called historical, but of materialistic investigations. These periods have been divided, to correspond with the materials from which their implements and weapons are made, viz., into the stone, the bronze, and the iron ages."

knowledge to the productive sphere. As Kuznets has stated: "The epochal innovation that distinguishes the modern economic epoch is the extended application of science to problems of economic production."[14] Our awareness of the importance of modern science (an awareness that is not, of course, confined to its purely economic significance) has led to a mushrooming interest in the history of science, which is now a thoroughly respectable academic discipline. But although we study the history of science (in some cases with the financial support of the National Science Foundation), the study of the history of technology is still largely (although by no means entirely) neglected. And yet to the extent that we are interested in the economic importance of science, we need to study the history of technology, because not all technologies will permit, or will permit in equal degrees, the *application* of scientific knowledge to the productive sphere. The growth of science, *by itself*, is not a sufficient condition for the growth of productivity. To believe that it is, is to ignore the mediating role of technology between man and nature. It was one of Marx's most important accomplishments to have posed precisely this question: What are the characteristics of technologies that make it possible to apply scientific knowledge to the productive sphere? Moreover, I think Marx suggested an answer that was quite adequate for his own historical period, given the nature of the industrial technology of his day and the state of scientific development. Still, this immensely important question needs to be posed and studied anew for the far more complex technologies, as well as the far more sophisticated bodies of scientific knowledge, that have emerged in the century since Marx wrote *Capital*. From the vantage point of the mid-nineteenth century, Marx's answer ran along the following lines.

The manufacturing system, which was the dominant mode of production of early capitalism, developed a high degree of worker specialization. Whereas the medieval handicraftsmen performed a whole range of operations in the production of a single commodity, the manufacturing system broke down the productive process into a series of discrete steps and assigned each step to a separate detail laborer. However, although this growing specialization of work had highly significant consequences with which Marx was intensely concerned, it nevertheless shared a basic feature in common with the medieval handicraft system: a continued reliance upon human skills and capacities.

> Whether complex or simple, each operation has to be done by hand, retains the character of a handicraft, and is therefore dependent upon the strength, skill, quickness, and sureness, of the individual workman

[14] Simon Kuznets, *Modern Economic Growth* (New Haven, 1966), p. 9.

> in handling his tools. The handicraft continues to be the basis. This narrow technical basis excludes a really scientific analysis of any definite process of industrial production, since it is still a condition that each detail process gone through by the product must be capable of being done by hand and of forming, in its way, a separate handicraft. It is just because handicraft skill continues, in this way, to be the foundation of the process of production, that each workman becomes exclusively assigned to a partial function, and that for the rest of his life, his labor power is turned into the organ of this detail function.[15]

Although, therefore, the manufacturing system achieved a growth in productivity through the exploitation of a new and more extensive division of labor, a rigid ceiling to the growth in productivity continued to be imposed by limitations of human strength, speed, and accuracy. Marx's point, indeed, is more general: Science itself can never be extensively applied to the productive process so long as that process continues to be dependent upon forces the behavior of which cannot be predicted and controlled with the strictest accuracy. Science, in other words, must incorporate its principles in impersonal machinery. Such machinery may be relied upon to behave in accordance with scientifically established physical relationships. Science, however, cannot be incorporated into technologies dominated by large-scale human interventions, for human action involves too much that is subjective and capricious. More generally, human beings have wills of their own and are therefore too refractory to constitute reliable, that is, controllable inputs in complex and interdependent productive processes.

The decisive step, then, was the development of a machine technology that was not heavily dependent upon human skills or volitions, where the productive process was broken down into a series of separately analyzable steps. The historic importance of the manufacturing system was that it had provided just such a breakdown. The historic importance of Modern Industry was that it incorporated these separate steps into machine processes to which scientific knowledge and principles could now be routinely applied. "The principle, carried out in the factory system, of analyzing the process of production into its constituent phases, and of solving the problems thus proposed by the application of mechanics, of chemistry, and of the whole range of the natural sciences, becomes the determining principle everywhere."[16] When this stage has been reached, Marx argues, technology becomes, for the first time, capable of indefinite improvement:

[15] *Capital,* Vol. 1, pp. 371–2.
[16] Ibid., p. 504.

> Modern industry rent the veil that concealed from men their own social process of production, and that turned the various, spontaneously divided branches of production into so many riddles, not only to outsiders, but even to the initiated. The principle which it pursued, of resolving each process into its constituent movements, without any regard to their possible execution by the hand of man, created the new modern science of technology. The varied, apparently unconnected, and petrified forms of the industrial processes now resolved themselves into so many conscious and systematic applications of natural science to the attainment of given useful effects. Technology also discovered the few main fundamental forms of motion, which despite the diversity of the instruments used, are necessarily taken by every productive action of the human body; just as the science of mechanics sees in the most complicated machinery nothing but the continual repetition of the simple mechanical powers.[17]

I suggest that Marx's insight into the historical interrelationships between science and technology was extraordinarily perceptive and that it ought to be treated as a starting point for the vastly more complex interrelationships that have characterized the last century of capitalist development.[18]

IV

In the remainder of this paper, I propose to concentrate upon Marx's analysis of the unique role and importance of the capital-goods sector. Although Marx is generally recognized as the father of the two-sector model, this recognition has been confined primarily to the usefulness of such models in the explanation of the inherent instability of capitalist economies, after the fashion in which Marx employed such models in the second volume of *Capital*. His work, however, suggests much more than this. The identification and isolation of a capital-goods-producing sector offers rich possibilities for the further understanding of the form and the mechanism of diffusion of technological change, as well as other critical aspects of the behavior of capitalist societies.

In the early stages of the development of Modern Industry, machin-

[17] Ibid., p. 532. For further discussion of these and related issues, see Nathan Rosenberg, "Karl Marx on the Economic Role of Science," *Journal of Political Economy*, July–August 1974, pp. 713–28.

[18] For some tentative exploration of these relationships, see Nathan Rosenberg, "Science, Invention and Economic Growth," *Economic Journal*, March 1974, pp. 90–108.

ery was produced by handicraft and manufacturing methods.[19] Although such methods were sufficient in the early stages, the growth in machine size and complexity and requirements for improvement in machine design and performance characteristics created demands that proved to be incompatible with the limited capacities of handicraft and manufacturing technologies. Indeed, "Such machines as the modern hydraulic press, modern powerloom, and the modern carding engine, could never have been furnished by Manufacture."[20] The realization of the full productive possibilities of Modern Industry therefore required that machine techniques be employed in the construction of the machines themselves. This is the final stage in the "bootstrap" operation, the stage by which Modern Industry completes its liberation from the constraints of the old technology. "Modern Industry had therefore itself to take in hand the machine, its characteristic instrument of production, and to construct machines by machines. It was not till it did this, that it built up for itself a fitting technical foundation, and stood on its own feet. Machinery, simultaneously with the increasing use of it, in the first decades of this century, appropriated, by degrees, the fabrication of machines proper."[21]

Once this stage of technological maturity has been attained, modern capitalism may be regarded as being in full possession of those extraordinary technological means that sharply distinguish it from all earlier stages in the development of man's productive capacities:

> So soon . . . as the factory system has gained a certain breadth of footing and a definite degree of maturity, and, especially, so soon as its technical basis, machinery, is itself produced by machinery; so soon as coal mining and iron mining, the metal industries, and the means of transport have been revolutionized; so soon, in short, as the general conditions requisite for production by the modern industrial system have been established, this mode of production acquires an elasticity, a capacity for sudden extension by leaps and bounds that finds no hindrance except in the supply of raw material and the disposition of the produce.[22]

[19] "As inventions increased in number, and the demand for the newly discovered machines grew larger, the machine-making industry split up, more and more, into numerous independent branches, and division of labor in these manufactures more and more developed. Here, then, we see in Manufacturing the immediate technical foundation of Modern Industry. Manufacturing produced the machinery, by means of which Modern Industry abolished the handicraft and manufacturing systems in those spheres of production that it first seized upon" (*Capital,* Vol. 1, p. 417).

[20] Ibid., p. 418.

[21] Ibid., p. 420. See also pp. 417–18 and 421. [22] Ibid., p. 492.

It was one of Marx's enduring accomplishments that he was among the first to perceive the inevitability of the trend toward bigness. This perception was, again, firmly rooted in his careful study of technological forces at work in mid-nineteenth-century British capitalism. Marx asserted in Chapter 25 of the first volume of *Capital* ("The General Law of Capitalist Accumulation") the decisive economic advantages of capitalist production on a large scale. The nature of these advantages is carefully categorized and analyzed, and numerous specific examples are presented in Chapter 5 of the third volume ("Economy in the Employment of Constant Capital").

When, as a result of the process of capital accumulation, the capitalist economy has acquired a sufficiently large complement of capital goods, and therefore also a well-defined sector devoted to the production of capital goods, the system at this stage acquires a new source of productive dynamism. First of all there are the opportunities, when the scale of production is sufficiently large, for the exploitation of what we have come to call indivisibilities:

> In a large factory with one or two central motors the cost of these motors does not increase in the same ratio as their horse power and, hence, their possible sphere of activity. The cost of the transmission equipment does not grow in the same ratio as the total number of working machines which it sets in motion. The frame of a machine does not become dearer in the same ratio as the mounting number of tools which it employs as its organs, etc. Furthermore, the concentration of means of production yields a saving on buildings of various kinds not only for the actual workshops, but also for storage, etc. The same applies to expenditures for fuel, lighting, etc. Other conditions of production remain the same, whether used by many or by few.[23]

Furthermore, when production takes place on a sufficiently large scale, it eventually becomes worthwhile to take steps to utilize waste, or by-product, materials. "The general requirements for the reemployment of these 'excretions' are: large quantities of such waste, such as are available only in large-scale production, improved machinery whereby materials, formerly useless in their prevailing form, are put into a state fit for new production; scientific progress, particularly chemistry, which reveals the useful properties of such waste."[24]

[23] Marx, *Capital,* Vol. 3, p. 79.

[24] Ibid., p. 100. Later Marx adds: "The most striking example of utilizing waste is furnished by the chemical industry. It utilizes not only its own waste, for which it finds new uses, but also that of many other industries. For instance, it converts the formerly almost useless gas-tar into aniline dyes, alizarin, and, more recently, even into drugs" (p. 102).

Most generally, the existence of a large stock of capital goods now provides powerful economic incentives for innovations of a capital-saving nature. That is to say, there now exist great opportunities for increasing the rate of profit "by reducing the value of the constant capital required for commodity production."[25] This involves not only the development of improved machinery – steam engines that deliver a greater amount of power with the same expenditure of capital and fuel – but also *technological change in the machinery-producing sector itself*. As a result of such improvements in the machine-building sector, "although the value of the fixed portion of constant capital increases continually with the development of labor on a large scale, it does not increase at the same rate."[26] At this most advanced stage of capitalist development, interindustry relationships come to play a most important role, because the rate of profit in one industry now becomes a function of labor productivity in another industry. "What the capitalist thus utilizes are the advantages of the entire system of the social division of labor. It is the development of the productive power of labor in its exterior department, in that department which supplies it with means of production, whereby the value of the constant capital employed by the capitalist is relatively lowered and consequently the rate of profit is raised."[27]

From an even broader perspective, the rate of profit may be increased by any capital-saving innovations, in whatever form. A large class of such innovations will therefore include all measures that reduce the turnover period of capital. From an economy-wide point of view, this has been precisely the effect of the communications revolution, one of the effects of which was a drastic reduction in the requirements for circulating capital. Marx deals with this source of capital-saving in Chapter 4 of the third volume of *Capital* ("The Effect of the Turnover on the Rate of Profit"):

> The chief means of reducing the time of circulation is improved communications. The last 50 years have brought about a revolution in this field, comparable only with the industrial revolution of the latter half of the eighteenth century. On land the macadamized road has been displaced by the railway, on sea the slow and irregular sailing vessel has been pushed into the background by the rapid and dependable steamboat line, and the entire globe is being girdled by telegraph wires. The Suez Canal has fully opened East Asia and Australia to steamer traffic. The time of circulation of a shipment of commodities to East

[25] Ibid., p. 80. [26] Ibid., p. 81. See also p. 84.
[27] Ibid., pp. 81–2.

> Asia, at least twelve months in 1847 . . . , has now been reduced to almost as many weeks. The two large centers of the crises of 1825–1857, America and India, have been brought from 70 to 90 percent nearer to the European industrial countries by this revolution in transport . . . The period of turnover of the total world commerce has been reduced to the same extent, and the efficacy of the capital involved in it has been more than doubled or trebled. It goes without saying that this has not been without effect on the rate of profit.[28]

All this from someone who is often described as treating the process of technological innovation as if it were purely a labor-saving phenomenon!

One further point that deserves to be made here is that Marx does not regard the reliance upon a new power source as the crucial element in the development of machines. Indeed, as he points out, in many of the early machines man was himself the prime mover. The distinctive element, as our earlier discussion has suggested, is the transfer of the control over the tool out of human hands.[29] This transfer of control involves a quantum leap forward, since "the number of tools that a machine can bring into play simultaneously is from the very first emancipated from the organic limits that hedge in the tools of a handicraftsman."[30] Furthermore, modern technology has even, *mirabile dictu,* invented a substitute for the human hand itself in the form of Henry Maudsley's slide rest.[31] This simple but ingenious device, as Marx perceptively notes, replaces not any particular tool "but the hand itself." In this sense the slide rest is a technological breakthrough fully comparable in importance to the steam engine.

In his discussion of technological change within the capital goods sector, Marx has many acute observations that are tossed out without further development. New inventions often contain inefficient design features at the outset because the inventor has not shaken himself totally free of an earlier technology that is being displaced and whose operating principles have been rendered irrelevant. Marx observes:

> To what an extent the old forms of the instruments of production influenced their new forms at first starting, is shown by, amongst other things, the most superficial comparison of the present powerloom with the old one, of the modern blowing apparatus of a blast-furnace with the first inefficient mechanical reproduction of the ordinary bellows, and perhaps more strikingly than in any other way, by the attempts before the invention of the present locomotive, to construct a locomo-

[28] Ibid., p. 71.

[29] "From the moment that the tool is taken from man, and fitted into a mechanism, a machine takes the place of a mere implement" (*Capital,* Vol. 1, p. 408).

[30] Ibid. [31] Ibid., p. 420.

> tive that actually had two feet, which after the fashion of a horse, it raised alternately from the ground. It is only after considerable development of the science of mechanics, and accumulated practical experience, that the form of a machine becomes settled entirely in accordance with mechanical principles, and emancipated from the traditional form of the tool that gave rise to it.[32]

It is a shame that posterity has been deprived of some sketch of the ill-fated two-footed locomotive! But, more seriously, the last sentence can be read as a striking anticipation of some of the central ideas of Abbott Payson Usher, probably the most careful twentieth-century student of the history of technology.[33] For Marx is insistent that technology has to be understood as a social process. The history of inventions is, most emphatically, not the history of inventors. Here, as in so many other realms, Marx's position cannot be understood without dealing with the basic methodological question: What is the most appropriate unit of analysis? His answer is that for questions pertaining to long-term changes in technology, the individual is not the appropriate unit. In this particular instance Marx is insisting that technological change cannot be adequately understood by examining the contributions of single individuals, even though he often acknowledges the noteworthy contributions of such individuals. Rather, one needs to examine the way in which larger social forces continually alter the focus of technological problems that require solutions. Within this framework one may then examine how the productive process has, in the past, shaped the development of scientific and technological knowledge and skills.[34] One is then in a position to explore the social process of problem formulation and eventual solution. In all of this, however, although individual human beings are, inevitably, the actors, the *dramatis personae* of the historical process, the actual unfolding of the plot turns upon the larger social forces that shape their actions. Within this larger framework it is possible to see the efforts and contributions of numerous individuals

[32] Ibid., p. 418.
[33] See Abbott Payson Usher, *A History of Mechanical Inventions*, rev. ed. (Cambridge, Mass., 1954), especially Chapter 4, "The Emergence of Novelty in Thought and Action."
[34] In all of this, the natural environment plays a critical role. "It is not the tropics with their luxuriant vegetation, but the temperate zone that is the mother country of capital. It is not the mere fertility of the soil, but the differentiation of the soil, the variety of its natural products, the changes of the seasons, which form the physical basis for the social division of labor, and which, by changes in the natural surroundings, spur man on to the multiplication of his wants, his capabilities, his means and modes of labor. It is the necessity of bringing a natural force under the control of society, of economizing, of appropriating or subduing it on a large scale by the work of man's hand, that first plays the decisive part in the history of industry" (*Capital*, pp. 563–4).

even though those twin bastions of individualism, the patent office and writers of history textbooks, require that the names of single individuals be written alongside the names of particular inventions. But what is really involved is a process of a cumulative accretion of useful knowledge, to which many people make essential contributions, even though the prizes and recognition are usually accorded to the one actor who happens to have been on the stage at a critical moment.

Usher would agree with Marx's stricture, quoted earlier, that "a critical history of technology would show how little any of the inventions of the eighteenth century are the work of a single individual." Usher's own analysis turns upon the study of problem formulation in dealing with technology, focusing especially upon the process that he describes as "the setting of the stage." It is the correct setting of the stage that makes possible the eventual act of insight leading to the solving of the problem (whether Watt's separate condenser or Bessemer's converter). As Usher points out: "Our analysis . . . redefines the question. It is not necessary to explain the final act of insight; the task now consists in explaining how the stage is set to suggest the solution of the perceived problem."[35] Moreover, this is not the end of the process, because the initial act of insight is likely to lead to crude and primitive solutions. "The setting of the stage leads directly to the act of insight by which the essential solution of the problem is found. But this does not bring the process to an end. Newly perceived relations must be thoroughly mastered, and effectively worked into the entire context of which they are a part. The solution must, therefore, be studied critically, understood in its fullness, and learned as a technique of thought or action. This final stage can be described as critical revision."[36]

I would regard the completion of Usher's stage of critical revision as bringing us to essentially the same point in the development of a technology that Marx had in mind in referring to the process by which "the form of a machine becomes settled entirely in accordance with mechanical principles, and emancipated from the traditional form of the tool that gave rise to it." Marx was aware of the regular need for something resembling Usher's final stage of critical revision and attached considerable importance to it. Indeed, there is implicit in Marx's analysis a kind of "life cycle" in the development of new techniques of production. New machines, when first introduced, are usually economically inefficient for two quite distinct reasons. First of all, the initial model has not yet had the opportunity to be subjected to a rigorous examination of its

[35] *A History of Mechanical Inventions*, p. 78.
[36] Ibid., p. 65.

operations from which methods for performance improvement can be expected to flow. Marx pointed to continual improvements

> which lower the use-value, and therefore the value, of existing machines, factory building, etc. This process has a particularly dire effect during the first period of newly introduced machinery, before it attains a certain stage stage of maturity, when it continually becomes antiquated before it has time to reproduce its own value. This is one of the reasons for the flagrant prolongation of the working time usual in such periods, for alternating day and night shifts, so that the value of the machine may be reproduced in a shorter time without having to place the figures for wear and tear too high. If, on the other hand, the short period in which the machinery is effective (its short life vis-à-vis the anticipated improvements) is not compensated in this manner, it gives up so much of its value to the product through moral depreciation that it cannot even compete with hand-labor.[37]

The early model, therefore, is recognized to have a short life expectancy – a high rate of "moral depreciation" – and this expectation that it will shortly be swept away by the competition of improved models is an ever-present consideration in the mind of the capitalist.

Eventually the new machine, having been subjected to a series of design improvements, assumes a relatively stabilized form, and at this point it becomes possible for the capital-goods sector to develop techniques for producing the machine more cheaply. This is where the capital-goods sector plays its critical role in the ongoing competitive process:

> After machinery, equipment of buildings, and fixed capital in general, attain a certain maturity, so that they remain unaltered for some length of time at least in their basic construction, there arises a similar depreciation due to improvements in the methods of reproducing this fixed capital. The value of the machinery, etc., falls in this case not so much because the machinery is rapidly crowded out and depreciated to a certain degree by new and more productive machinery, etc., but because it can be reproduced more cheaply. This is one of the reasons why large enterprises frequently do not flourish until they pass into other hands, i.e., after their first proprietors have been bankrupted, and their successors, who buy them cheaply, therefore begin from the outset with a smaller outlay of capital.[38]

Marx emphasized in several other places the high cost of the early machine models by comparison with the later ones.[39] His argument

[37] *Capital,* Vol. 3, p. 112. [38] Ibid.

[39] Ibid., p. 103, and *Capital,* Vol. 1, p. 442. Marx cites Babbage for supporting evidence in both places and Ure in the former. Both arguments – with respect to improvements in

suggests a great deal, not only about the process through which a capitalist economy generates new techniques, but also about the speed with which new techniques will be spread throughout the economy.[40]

I do not propose to discuss the question of whether Usher's work was influenced by Marx. That is not, in my view, terribly important. But I do want to insist upon the fruitfulness of certain ways of conceptualizing the technological process under capitalism that Marx suggested but never developed beyond some precocious and suggestive hints. I want also to render my judgment that American students of Marx, in creating a mode of analysis that they call "the Marxist tradition," have not been faithful, in at least one important respect, to the mode of analysis initiated by Marx himself. For Marx, as I have argued, was a close student of both the history of technology and its newly emerging forms.

These strictures are only somewhat less applicable to the British Marxian tradition. For although it is true that the two leading British figures in the history of technology, J. D. Bernal and Joseph Needham, have frequently adverted to the strong Marxian component of their thinking, they have not, any more than others, followed Marx's hints or elaborated upon his insights concerning the development of modern industrial technology. This is a task that still remains to be undertaken.

machine design and improvements in techniques for producing the machines – are combined in *Capital*, Vol. 1, p. 442: "When machinery is first introduced into an industry, new methods of reproducing it more cheaply follow blow upon blow, and so do improvements, that not only affect individual parts and details of the machine, but its entire build. It is, therefore, in the early days of the life of machinery that this special incentive to the prolongation of the working day makes itself felt most acutely."

40 See Nathan Rosenberg, "Factors Affecting the Diffusion of Technology," *Explorations in Economic History*, Fall 1972, pp. 3–33.

PART II

Some significant characteristics of technologies

3 Technological interdependence in the American economy

One of the things that all knowledgeable people supposedly "know" is that technological change has been the critical variable in accounting for the spectacular long-term growth of the American economy and our resulting present affluence. And yet, when scholars of a quantitative turn of mind have attempted to link the story of the growing productivity of the American economy to some of the better-known facts and landmarks of our technological history, that story has turned out to be a remarkably difficult one to tell.

There are many reasons why this has been a difficult exercise. It is, for one thing, an extremely complicated methodological matter to separate out the contribution of technological change from other changes in human behavior, motivation, and social organization. Although this is generally realized, there is less awareness that the productivity contribution of a new technology is also linked to other, less obvious technological forces, to which I will shortly return. Moreover, the public image of technology has been decisively shaped by popular writers who have been mesmerized by the dramatic story of a small number of major inventions – steam engines, cotton gins, railroads, automobiles, penicillin, radios, computers, and so on. In addition, in the telling of the story, overwhelming emphasis is placed on the specific sequence of events leading up to the decisive actions of a single individual. Indeed, not only our patent law but also our history textbooks and even our language all conspire in insuring that a single name and date are attached to each invention.

The growing interest in the diffusion of technology in recent years has functioned as a partial corrective to the heroic theory of invention. Inventions acquire their economic importance, obviously, only as a function of their introduction and widespread diffusion. But I want to

This paper was originally published in *Technology and Culture,* January 1979, pp. 25–50.

go further and suggest that the social and economic history of technology can only be properly written by people possessing a close familiarity with the actual technology itself. At the same time, as I would also insist, people who know little beyond the technology in a narrow sense are not likely to rise above the level of antiquarianism. Indeed, one of the main themes of this paper is that it is absolutely essential not to develop too narrow a focus in the study of technology, because a narrow focus severs the links between a given technology and many of the factors that will, inevitably, determine its effectiveness and significance. A larger purpose in what follows will be to show how our appreciation for the functioning of technology in the growth of the American economy can be expanded by focusing attention on the network of larger technological relationships in which specific inventions are always embedded.

I do not want to concern myself here with the conceptual, methodological, and statistical problems involved in attempting to quantify the contributions of technological change to long-term economic growth in America. These problems have been extensively discussed elsewhere. Rather, I want to confine myself to certain aspects of the problem that are of greatest interest to an audience that is committed to the subject of the history of technology. Specifically, I would like to concentrate on certain intrinsic characteristics of the process of technological change that, I will argue, are central to the difficulties experienced when we attempt to measure the growth in productivity flowing from it. The central theme, on which I wish to elaborate, is that technological improvement not only enters the structure of the economy through the main entrance, as when it takes the highly visible form of major patentable technological breakthroughs, but that it also employs numerous and less visible side and rear entrances where its arrival is unobtrusive, unannounced, unobserved, and uncelebrated. It is the persistent failure to observe the rush of activity through these other entrances that accounts for much of the difficulty in achieving a closer historical linkage between technological history and the story of productivity growth. I will briefly explore these neglected factors under three main headings.

Complementarities

Inventions hardly ever function in isolation. Time and again in the history of American technology, it has happened that the productivity of a given invention has turned on the question of the availability of complementary technologies. Often these technologies did not initially exist, so that the benefits potentially flowing from invention A had to await the achievement of inventions B, C, or D. These relationships of complementarity therefore make it exceedingly difficult to predict the flow of benefits from any single invention and commonly lead to a postponement in the flow of such expected benefits. Technologies depend upon one another and interact with one another in ways that are not apparent to the casual observer, and often not to the specialist.

A serious difficulty in tracing out the social payoff to invention is that these linkages are both numerous and of varying degrees of importance and therefore difficult to measure with any pretense of precision. Thus an invention reducing the cost of power generation differentially affects different industries. In the past such cost reductions were critical to the expansion of the aluminum industry, an intensive user of electricity. They played a major role in the cheapening of commercial fertilizers and the increasing intensity with which such fertilizers were used in food production. Their significance for the production of ballpoint pens or umbrellas was probably very small. In the event that innovations in power generation were to bring about a massive reduction in power cost (a consummation devoutly to be wished!) further innovations, which are known to be technically feasible but economically unattractive at present, might move into the realm of economic feasibility (e.g., various methods of desalination).

Consider, alternatively, the economic payoff to an innovation that reduced the cost of transportation, for example, the impact of the railroad on the American economy in the mid-nineteenth century.[1] Part of the economic payoff consisted of an increase in the productivity of agriculture, as bulky farm products could now be exchanged over larger geographic areas than was formerly possible. As a consequence it became possible to engage in a greater degree of regional specialization than before and to participate more fully in the increased productivity resulting from the improved opportunities for devoting heterogeneous agricultural resources to their best possible uses. Any estimate of the social payoff from the railroad would need to include an estimate of the reduction in transport costs attributable to it, the effect of cheaper transport on possibilities for regional specialization, and the size of the benefits specifically attributable to this increased specialization, a specialization that makes possible a much finer adaptation of the productive process to the geographic distribution of resources. Furthermore, reductions in transport costs also bring about greater productivity by making it possible to concentrate output in a smaller number of more

[1] See Albert Fishlow, *American Railroads and the Transformation of the Ante-Bellum Economy* (Cambridge, Mass., 1965); and Robert W. Fogel, *Railroads and American Economic Growth* (Baltimore, 1964).

efficient units. For example, reductions in the cost of transporting coal from the mine to electric generating facilities, due to the development of the unitized train shipment of coal and other transportation improvements,[2] have made it possible to close down less efficient coal mines that had previously survived because of their proximity to markets and to rely increasingly on a smaller number of more efficient mining operations. Finally, reductions in transport costs generally make possible the more intensive exploitation of economies of scale, wherever these economies may be significant.[3] In a world of high transport costs, the size of operation of an individual plant will be constrained by the prohibitively high cost of transport as the product is moved to more distant markets. Reductions in transport costs expand the market available to a firm in any given location and thus increase the possibilities for the exploitation of scale economies.

On an even wider geographic scale, the social payoff resulting from railroads and associated reductions in transport cost was increased by the iron steamship, which reduced the cost of transoceanic shipping, and refrigeration, which in turn raised the productivity of *both* the railroad and the steamship. With these complementary innovations there began to emerge, by the end of the nineteenth century, a truly worldwide agricultural division of labor.[4] This unique division of labor was the combined result of reductions in the cost of land and water transport and the newly acquired ability to utilize the railroad and the steamship for the long-distance shipment of meat. By the 1880s and 1890s, as a result of refrigeration techniques, the rapidly growing populations of western Europe were becoming heavily dependent upon a wide range of overseas food products, including not only the North American midwest but also large quantities of lamb from New Zealand and Australia and beef from the Argentine.[5]

This emphasis on complementarities serves to make explicit one of the main points of this paper: The social payoff of an innovation can rarely be identified in isolation. The growing productivity of industrial

[2] U.S. Department of Labor, *Technological Trends in Major American Industries* (Washington, D.C., 1966), p. 20.

[3] Innovations leading to the reduction in high-voltage transmission costs have exactly the same effect. They make it possible to shut down relatively small, older plants and to exploit the economies of large-scale power generation in a limited number of localities.

[4] By 1903 "freight rates in general were down to about 20 percent of the 1877–8 level, and the actual costs of ocean shipment had fallen by an even greater percentage due to reductions in the cost of insurance" (A. J. Youngson, "The Opening Up of New Territories," in *The Cambridge Economic History of Europe*, vol. 6, *The Industrial Revolutions and After*, ed. H. J. Habakkuk and M. Postan [Cambridge, 1965], pt. 1, p. 171).

[5] Ibid., pp. 172–3.

economies is the complex outcome of large numbers of interlocking, mutually reinforcing technologies, the individual components of which are of very limited economic consequence by themselves. The smallest relevant unit of observation, therefore, is seldom a single innovation but, more typically, an interrelated clustering of innovations. The early industrial revolution can only be understood in terms of the interactions of a few basic technologies that provided the essential foundation for other technological changes in a series of ever-widening concentric circles, at the heart of which were a few major innovations in steam power, metallurgy (primarily iron), and the large-scale utilization of mineral fuels. One can identify similar kinds of clusterings around electrification beginning in the late nineteenth century, the internal combustion engine in the early twentieth century, and plastics, electronics, and the computer in more recent years. In each case a central innovation, or small number of innovations, provided the basis around which a larger number of further cumulative improvements and complementary inventions were eventually positioned.

The importance of these complementarities suggests that it may be fruitful to think of each of these major clusterings of innovations from a systems perspective. The systems nature of a body of technology is well displayed in the case of the electric light. Indeed, it is clear that the most successful inventor-innovators in the development of the electric light were successful – in good part at least – because they consciously and deliberately approached the industry from a systems framework. Incandescent lighting constituted a system of several major components. Economic success in the innovation process was contingent on considering all aspects of the system in the delivery of light to domestic residences. Many of the numerous instances of entrepreneurial failure can be attributed to the fact that a would-be entrepreneur failed to consider the relevant conditions of interdependence between the component with which he happened to be preoccupied and the rest of the larger system.[6] This system can be considered as consisting of four significant components: (1) the generation of electricity at a central power station, (2) a conductor network for the transmission of power, (3) a meter to measure household consumption of electricity, and (4) a lamp. Successful inventor-innovators in incandescent lighting, such as Thomas A. Edison

[6] Passer points out, for example, that "the relatively poor performance of the United States [Electric Lighting] company can be partly attributed to the fact that its technical personnel, while competent, did not realize the importance of developing an entire incandescent-lighting system, rather than certain components" (Harold C. Passer, *The Electrical Manufacturers. 1875–1900* [Cambridge, Mass., 1953], p. 148, emphasis Passer's; see also pp. 176–7).

and George Westinghouse, consciously thought in terms of the entire system, the purpose of which was to deliver cheap illumination into millions of domestic residences. "Both set out to develop an entire system, and both took a personal interest in the invention of the system components."[7] In Edison's case:

> The parallel between Edison's work on the dynamo and on the incandescent light is apparent. In each case, he perceived the function of the component in the system. He then determined the characteristic of the component which would result in a system with the lowest production cost of light. The next step was to apply the electrical principles and to conduct numerous experiments until the desired end was reached.
>
> The transmission network which connected the dynamo and the lamps was a third main component of the Edison lighting system. As in the lamp and the dynamo, Edison's contribution was to invent a cost-reducing component.[8]

It is characteristic of a system that improvements in performance in one part are of limited significance without simultaneous improvements in other parts, just as the auditory benefits of a high-quality amplifier are lost when it is connected to a hi-fi set with a low-quality loudspeaker. (For example, after the introduction of steel rails made possible the use of longer trains with heavier loads traveling at higher speeds, making them much more difficult to stop, Westinghouse "providentially" developed the air brake. The improved design of automobile engines and greater speeds were likely to be disastrous without a better braking system and better engineered roads.) Similarly, improvements in power generation will have only a limited impact on the delivered cost of electricity until improvements are made in the transmission network and the cost of transporting electricity over long distances. This need for further innovations in complementary activities is an important reason why even apparently spectacular breakthroughs usually have only a gradually rising productivity curve flowing from them. Really major improvements in productivity therefore seldom flow from single technological innovations, however significant they may appear

[7] Ibid., p. 192; see also Arthur A. Bright, Jr., *The Electric Lamp Industry* (New York, 1949), pp. 67–9, 76.

[8] Passer, pp. 177–8. See also the subsequent discussion of the fourth component, the meter. The meter was an extremely important component in the system. Before General Electric developed a satisfactory meter around 1900, meters were likely to be both very expensive and unreliable, and flat-rate contracts were common. Consumers had no incentive to economize in the use of electricity. In the absence of a meter, therefore, electrical utilities had to undertake excessively large investments in generating and transmitting equipment, and operating expenses were very high. The meter, therefore, was a major contributor to the improved efficiency in resource use.

to be. But the combined effects of large numbers of improvements within a technological system may be immense.[9] Moreover, there are internal pressures within such systems that serve to provide inducement mechanisms of a dynamic sort. One invention sharply raises the economic payoff to the introduction of another invention. The attention and effort of skilled engineering personnel are forcefully focused on specific problems by the shifting succession of bottlenecks that emerge as output expands.[10]

The role of complementarity relationships may be further observed, in finer detail, in the history of individual innovations. Sometimes a particular innovation has to await the availability of a specific complementary input or component; sometimes the evident need for the input is sufficient to lead to its invention; and sometimes the input, when it is fully developed, is found to have uses and applications of a totally unanticipated – or at least unintended – sort. Thus, many innovations have had to await the development of appropriate metallurgical inputs with highly specific performance characteristics. The compound steam engine had to await cheap, high-quality steel. Higher pressures (and therefore greater fuel economy) in power generation required high-strength, heat-resistant alloy steels.[11] Hard alloy steels, in turn, were of limited usefulness until appropriate new machine tooling methods were developed for working them. The jet engine required, and eventually contributed to, numerous metallurgical improvements. Similarly, the transistor required major improvements in techniques for purifying metals and eventually contributed richly to a wide range of productive activities that also required metals of a high degree of purity. In agriculture, the introduction of techniques for the mechanical harvesting of crops has been sharply accelerated by the advances in genetic knowl-

[9] Bright cites an estimate for the reduction in residential lighting costs that "takes into account the reductions in energy cost, the reductions in lamp price, the increases in lamp efficiency, and the increase, if any, in lamp life . . . Lighting costs in 1945 were 1.3 percent of what they were in 1882; they were 13 percent of what they were in 1906; and they were 45 percent of what they were in 1923. About 60 percent of the saving since 1923 is attributable to increases in lamp efficiency; and about 10 percent is attributable to reductions in lamp prices" (p. 362).

[10] For further discussion of the role of bottlenecks in inducing technological changes, see Nathan Rosenberg, "The Direction of Technological Change," *Economic Development and Cultural Change* 17 (October 1969): 1–24.

[11] On high-pressure steam engines, Usher stated: "Undoubtedly, the limiting factor was not the concept, but the practical difficulty of dealing with steam pressures. Neither boilers nor cylinders could then be made that would resist the pressures needed for effective working" (A. P. Usher, *A History of Mechanical Inventions* [Boston, 1959], p. 356).

edge that permit a redesigning of the plant itself to accommodate the specific needs of machine handling. Thus, midwestern corn is now almost entirely of a specially bred, stiff-stalked variety that remains conveniently upright well into the fall; mechanical tomato harvesters have been available for many years but were not adopted until it became possible to breed a new, tough-skinned variety that was less susceptible to bruising and ripened more uniformly; similarly, in cotton, breeding has been directed toward the development of plants that lend themselves more readily to mechanical picking.[12]

The cumulative impact of small improvements

I turn now to a second significant aspect of technology. That is, a large portion of the total growth in productivity takes the form of a slow and often almost invisible accretion of individually small improvements in innovations. The difficulty in perception seems to be due to a variety of causes: to the small size of individual improvements; to a frequent preoccupation with what is *technologically* spectacular rather than *economically* significant; and to the inevitable, related difficulty that an outsider has in attempting to appreciate the significance of alterations within highly complex and elaborately differentiated technologies, especially when these alterations are, individually, not very large.

It is useful here to think in terms of the life cycle of individual innovations. Major improvements in productivity often continue to come long after the initial innovation as the product goes through innumerable minor modifications and alternations in design to meet the needs of specialized users.[13] Widely used products such as the steam engine or the electric motor or the machine tool experience a proliferation of changes as they are adapted to the varying range of needs of ultimate users. Consumer durables have typically gone through parallel experiences with special emphasis on expanding the quality range in catering to different income categories.[14] Such modifications are achieved by unspectacular design and engineering activities, but they constitute the substance of much productivity improvement and increased consumer well-being in industrial economies.

[12] See Clarence Kelly, "Mechanical Harvesting," *Scientific American* (August 1967), pp. 50–9; Wayne Rasmussen, "Advances in American Agriculture: The Mechanical Tomato Harvester as a Case Study," *Technology and Culture* 9 (October 1968): 531–43.

[13] Marx pointed out that there were no less than 500 different types of hammers being produced in Birmingham (Karl Marx, *Capital* [New York, n.d.], p. 375).

[14] Brady provides extensive documentation for individual products (see Dorothy Brady, "Relative Prices in the Nineteenth Century," *Journal of Economic History* [June 1964], pp. 146–7, 155–6, 164, 75–82, and passim).

The view of technological change as consisting of steady cumulation of innumerable minor improvements and modifications, with only very infrequent major innovations, was nicely embodied by S.C. Gilfillan in his book, *Inventing the Ship*.[15] Although Gilfillan was primarily concerned with the social rather than the economic aspects of the process, his book provides an invaluable "close-up" view of the gradual and piecemeal nature of technological change, drawing heavily on small refinements based on experience and gradually incorporating a succession of improved components or materials developed in other industries. His analysis of the evolution of marine engines (Chap. 2) is that of a slow sequence incorporating the growing strength and steam-raising capacity of boilers, the increasing reliance on steel components as steel became cheaper, and the adoption of petroleum lubricants:

> To the ship's motive plant were added further important cut-off arrangements and valve gear for them, feed-water heaters (e.g., from the condenser), superheaters (in the 60s, saving 10 percent of the fuel), steam jackets, better air-pumps, evaporators, tricks of tinning the copper condenser tubes, changing to brass after learning how to manufacture the brass, protecting the tubes from galvanic action, forced draft, and various improvements of grates and methods of feeding with coal and air, instead of iron (a large improvement for weight-saving), and a limitless number of betterments too minor for us to mention here.[16]

The introduction of screw propulsion (Chap. 3) was largely a matter of determining, through experience and experiment, the optimal design form of the propeller as well as simply exploiting new construction possibilities provided by improvements in metalworking as they occurred elsewhere: "The propeller admits of strangely wide variations in form without much difference of efficiency, if only certain gradually learned mathematical principles be respected, chiefly that of adapting the blade angle at every separate point to the speeds of ship, engine and thrown water. It was in learning such principles that the real invention of the propeller largely took place."[17] Shipbuilding for the past century has been involved in a long sequence of gradual improvements: improvements in engine efficiency that save fuel space; changes in hull design; exploitation of scale economies that permit reductions in crew requirements per ton of cargo; changes in cargo handling techniques,

[15] S. C. Gilfillan, *Inventing the Ship* (Chicago, 1935). See also his companion volume, *The Sociology of Invention* (Chicago, 1935), and A. P. Usher's earlier and authoritative *A History of Mechanical Inventions*.
[16] Gilfillan, *Inventing the Ship*, p. 131. [17] Ibid., p. 137.

such as containerization, that also sharply reduce "turn-around time," and so on.[18]

Louis Hunter's observations on the history of the steamboat on western rivers in the antebellum period are worth quoting here because his description of technological change in connection with the steamboat would also apply, with only minor changes, to a broad range of technological change elsewhere and in later periods.

> The history of the steamboat is also the history of foundry and machine-shop practice, of metalworking techniques and machine tools, and of the practical art of steam engineering. The story is not, for the most part, one enlivened by great feats of creative genius, by startling inventions or revolutionary ideas. Rather, it is one of plodding progress in which invention in the formal sense counted far less than a multitude of minor improvements, adjustments, and adaptations. The heroes of the piece were not so much such men as Watt, Nasmyth, and Maudslay, Fulton, Evans, and Shreve – although the role of such men was important – but the anonymous and unheroic craftsmen, shop foremen, and master mechanics in whose hands rested the daily job of making things go and making them go a little better. The story of the evolution of steamboat machinery in the end resolves itself in large part into such seemingly small matters as, for instance, machining a shaft to hundredths instead of sixteenths of an inch, or devising a cylinder packing which would increase the effective pressure a few pounds, or altering the design of a boiler so that cleaning could be accomplished in three hours instead of six and would be necessary only every other instead of every trip. Matters such as these do not often get into the historical record, yet they are the stuff of which mechanical progress is made, and they cannot be ignored simply because we know so little about them.[19]

Much of the technological change that goes on in an advanced industrial economy is, if not invisible, at least of a low-visibility sort. It includes a flow of improvements in materials handling,[20] redesigning

[18] "Container ships may reduce terminal loading costs by 90 percent and 'turn-around time' from 84 to 13 hours. The gang that was capable of loading 25 tons 'loose stow' in 1 hour can load 300 tons of containerized cargo in the same time. Use of pallets and containers increases productivity by 3 to 4 times." (U.S. Department of Labor, *Technological Trends in 36 Major American Industries* [Washington, D.C., 1964], p. 78.)

[19] Louis Hunter, *Steamboats on the Western Rivers* (Cambridge, Mass., 1949), pp. 121–2.

[20] For example, in the construction industry "there are a plethora of materials handling improvements. They range from hoists of all types, to conveyors, to higher line speeds, to powered concrete buggies, to more handleable packages on the part of suppliers. These improvements have been continuous and probably no single change is individually significant. We do have an estimate of the use of one type, the tower crane, which has changed

production techniques for greater convenience, and reducing maintenance and repair costs (as in modular machinery design).[21] In iron and steel, reductions in fuel requirements have been achieved by rearrangement of plants so as to eliminate the need for successive reheating of materials.[22] In metalworking, new and harder materials continue to be introduced in cutting edges, making possible a considerable acceleration in the pace of work.[23] In electric-power generation, where the long-term rate of growth of total factor productivity has been higher than in any other American industry,[24] the slow, cumulative improvements in the efficiency of centralized thermal power plants have generated enormous long-term increases in fuel economy. A stream of minor plant improvements, including the steady rise in operating temperatures and pressures made possible by metallurgical improvements (such as new alloy steels) and the increasing sophistication of boiler design and resulting increased capacity, have sharply raised energy output per unit of input. The size of this improvement may be indicated as follows. It required almost 7 pounds of coal to generate a kilowatt-hour of electricity in 1900, but the same amount of electricity could be generated by less than 0.9 pound of coal in the 1960s.[25] But even this figure understates

the construction skyline in recent years. Not used in this country in 1958, 150 European tower cranes had been imported by 1962 and the number [is] increasing" (A. D. Little, Inc., *Patterns and Problems of Technical Innovation in American Industry* [report to National Science Foundation, September 1963], p. 132).

[21] Innumerable such examples may be found in U.S. Department of Labor, *Technological Trends in 36 Major American Industries* (1964), and *Technological Trends in Major American Industries,* Bulletin no. 1474 (Washington, D.C., 1966).

[22] The reduction in fuel requirements from all sources – a trend going back well into the nineteenth century for the steel industry – shows no sign of abating. In 1949 it required almost 1,900 pounds of coke to produce a ton of pig iron; in 1968 it required only 1,200 pounds (see Bureau of Mines, Department of Interior, *Mineral Facts and Problems* [Washington, D.C., 1970], p. 40).

[23] "From 1935 to 1955 the machine tool industry made rapid progress in increasing metal-cutting speeds. Whereas in 1935 cutting speeds were 150–200 feet per minute with high-speed tools, by 1955 they had reached 600–800 feet per minute with carbide tools and more than 1,000 feet per minute with ceramic tools. The effect of speed on the cost of cutting is easily calculable; doubling the speed halves the time and the cost. However, there is another consideration. Tool life decreases with increased speed; consequently, tool maintenance and replacement costs increase. The most economical cutting speed, for a given set of conditions, therefore, represents a compromise between the two rates . . . The industry was able to reduce the productive cost of metal cutting by as much as 75% during this 20 year period" (Little, p. 99).

[24] John Kendrick, *Productivity Trends in the United States* (Princeton, N.J., 1961), pp. 136–7.

[25] Hans H. Landsberg and Sam H. Schurr, *Energy in the United States* (New York, 1968), pp. 60–1.

the full improvement in the utilization of energy sources: "During the 50-year period 1907–1957 reduction of the total energy required or lost in coal mining, in moving the coal from mine to point of utilization, in converting to electric energy, in delivering the electric energy to consumers, and in converting electric energy to end uses have increased by well over 10 times the energy needs supplied by a ton of coal as a natural resource."[26]

In the construction industry, often regarded as a stronghold of traditionalism and conservatism, there have been innumerable minor changes of great cumulative significance, but it may be that the organizational changes have been even more significant:

> During the last thirty years, the U.S. building industry has undergone a radical change of character. Project and corporate size has increased greatly. Equipment, materials, design and planning practices are in many ways different than those employed before the Depression. Nevertheless, while the industry as a whole has undergone major change, this change has proceeded in the small segments of the industry through many small increments. There has been no radical change, of great technical and economic significance, which is associated with a single invention or family of inventions. Nothing is to the building industry as synthetic fibers and finishes are to textiles or as numerical controls are to machine tools. In the building industry, change has been evolutionary – like the many small process changes accounting for increased productivity in machine tools and textiles – and much of the most important change cannot be described as technical at all. It has had to do, rather, with methods of managing and organizing the building process.[27]

A more general source of small, low-visibility innovations of great cumulative significance has been the multitude of ways in which main-

[26] U.S. Department of Commerce, *Historical Statistics of the United States* (Washington, D.C., 1960), p. 501. See also William Hughes, "Scale Frontiers in Electric Power," in *Technological Change in Regulated Industries,* ed., William Capron (Washington, D.C., 1971).

[27] Little, p. 119. The availability of superior materials has been particularly important to construction: "Improvements in construction materials make possible more efficient utilization. Paints, for example, require less on-site preparation and less effort in their application. Adhesives are being more widely used to save time and reduce wall costs. Plastics offer the advantage of ease of handling and ability to be molded to extremely close tolerances. The development of high-strength and rust-retardant steels allows construction in which the steel is exposed to the weather. Labor and other cost savings of 25 percent can be realized by the use of prestressed concrete beams in place of structural steel in some areas. Prestressed concrete also makes possible wide spans where column-free contruction is desirable. Brick construction has benefited by the development of high-strength mortar" (U.S. Department of Labor, 1964, pp. 12–15).

tenance and service requirements for capital goods have been reduced and the useful life of capital goods prolonged. The substitution of new materials (e.g., aluminum and rust-resistant steels) for old ones and improved techniques of friction reduction (lubrication and roller bearings) have led to a considerable extension of the useful life of a wide range of capital equipment. The replacement of untreated railroad ties with ties impregnated with creosote was estimated roughly to double the expected life of a tie – from fourteen to twenty-eight years. Sludge removers and chemically treated feed water extended the life of locomotive boilers and reduced the frequency with which they once had to be taken out of service and washed out.[28] The substitution of heavier for lighter rails increased the life of a rail by a percentage far in excess of the weight increase.

The emphasis placed on technological change of the form emphasized here suggests the extremely great usefulness of research that attempts to link productivity change with specific technological changes at the level of the individual firm. Ideally, studies conducted at the firm level, and with sufficient access to appropriate technological and economic information, should be able to accomplish what has not been done at the highly aggregated level: to separate out the contribution of technological changes from the variety of other forces contributing to the growth of productivity. One such microeconomic analysis is Samuel Hollander's study of the du Pont rayon plants,[29] in which he attempts to determine the extent to which observed reductions in unit costs of production at particular plants are the result of changes in the techniques of production. Hollander's findings are of great interest in the present context. Unit costs declined strikingly in the du Pont plants which he studied. Furthermore, he finds that the contribution of technical change in accounting for these reductions was "of overwhelming

[28] Harold Barger, *The Transportation Industries*, 1889–1946 (New York, 1951), pp. 100–11. For some useful estimates of the impact of improved maintenance procedures on employment requirements for American railroads, see William Haber et al., *Maintenance of Way Employment on U.S. Railroads* (Detroit, 1957). The story of the response of the railroads to rising timber prices is told in Sherry Olson, *The Depletion Myth* (Cambridge, Mass., 1971). It is interesting to note that chemical techniques for the preservation of wood also had the important result of making it possible to use "inferior species" of wood for crossties – kinds of wood that decayed very rapidly without chemical treatment and that were not used until chemical treatment became widespread. In this respect such techniques not only increased the useful life of crossties but expanded substantially the wood supplies upon which it became possible to draw. "Untreated, the mixed hardwoods, sappy pines, and Douglas fir had small value as ties or bridge timber. Treated, their service value for ties and many items of car lumber was roughly equal to the best white oak" (ibid., p. 132).

[29] Samuel Hollander, *The Sources of Increased Efficiency: The Study of Du Pont Rayon Plants* (Cambridge, Mass., 1965).

importance."[30] And, most significant for our present purposes, is his finding that the cumulative effect of minor technical changes on cost reduction was actually greater than the effect of major technical changes.[31]

Hollander is, of course, aware that there is an interdependence between minor and major technical changes and that "without some preceding major change the potential stream of minor changes will be exhausted."[32] Nevertheless, his findings lend powerful support to the view that the economic importance of minor technical improvements has been vastly underestimated.

Hollander's findings for rayon are closely paralleled by those of Enos in his study of technological change in petroleum refining. Enos studied the introduction of four major new processes in petroleum refining: thermal cracking, polymerization, catalytic cracking, and catalytic reforming. In measuring the benefits for each new process he distinguished between the "alpha phase" – or cost reductions that occur when the new process is introduced – and the "beta phase" – or cost reductions that flow from the later improvements in the new process. Enos found that the average annual cost reductions that were generated by the beta phase of each of these innovations considerably exceeded the average annual cost reductions that were generated by the alpha phase (4.5 percent as compared with 1.5 percent). On this basis he asserted that "the evidence from the petroleum refining industry indicates that improving a process contributes even more to technological progress than does its initial development."[33]

[30] Ibid., pp. 192–3. [31] Ibid., p. 196. [32] Ibid., 205.

[33] John L. Enos, "A Measure of the Rate of Technological Progress in the Petroleum Refining Industry," *Journal of Industrial Economics* (June 1958), p. 180. In their study of the operation of the steamboat on western rivers in the antebellum period, James Mak and Gary Walton also emphasize the quantitative importance of later improvements in the steamboat as compared to the cost reductions that had been achieved by the initial innovation. Thus "The introduction of the steamboat, 1815–20, led to a significant fall in real freight costs, but the absolute as well as the relative decline in real freight rates was greatest during the period of improvement, 1820–60" (James Mak and Gary Walton, "Steamboats and the Great Productivity Surge in River Transportation," *Journal of Economic History* [September 1972], p. 625). Not all of the improvement in productivity, of course, was attributable to technological change. For example, significant reductions in cargo collection times and passage times were unconnected to such changes. However, technological changes brought about increases in cargo-carrying capacity per measured ton and an extension of the navigation season. A cumulation of minor design changes on the steamboat had the effect of substantially increasing the length of the navigation season for each steamboat size class. By steadily reducing the draft in relation to tonnage and cargo-carrying capacity, steamboat designers and builders brought about major improvements in the productivity of capital by enabling steamboats to operate a longer portion of

Albert Fishlow's incisive study of productivity growth and technological change in the railroad sector makes a notable contribution at the industry level to the goal of sorting out the relative importance of the separate factors contributing to that growth.[34] Fishlow calculates the growth in productivity of American railroads between 1870 and 1910. That growth was extremely large. He finds that the incremental expenses that would have been required to meet the demands of 1910 traffic loads with the technology available back in 1870 would have amounted to about $1.3 billion. The technological sources of productivity growth included a series of important inventions specific to the railroads – air brakes and automatic couplers – the substitution of steel for iron rails, and the gradual improvement in the design of locomotives and rolling stock. Fishlow finds that the economic contribution of the air brake and the automatic coupler was minor. The higher speed and greater safety due to these inventions translated into operational economies of $50 million. The substitution of steel rails for iron, however, was of major importance, and such rails were rapidly adopted in spite of their much higher price. Steel rails were first used by the Pennsylvania Railroad during the Civil War, and they accounted for 80 percent of all track mileage by 1890. Steel rails were far more durable, lasting more than ten times as long as iron rails, and they could bear far greater loads than iron rails. Indeed, the old iron rails of 1870 were simply incapable of supporting the 1910 locomotives and would have been crushed under their average weight of 70 tons. Fishlow calculates that the combined effects of increased longevity and greater strength effected a saving of $479 million.

But the largest cost saving by far was due to a succession of improvements in the design of locomotives and freight cars, even though the process included no readily distinguishable or memorable innovations.

the year. In fact, as a rough average, "The navigation season was extended from approximately six months, before 1830, to about nine months, during the last half of the antebellum period" (ibid., p. 634). Here, too, the greatest overall increase in total factor productivity came in the years following the initial introduction of the innovation. According to Mak and Walton, "The major factor causing the reduction of input requirements per payload ton was the more than threefold increase in the ratio of carrying capacity to measured tonnage. If utilization and the ratio of carrying capacity to measured tonnage had remained unchanged over the period, we would have observed little change in the input requirements of capital, labor, and fuel per payload ton (where we consider only a single voyage)" (p. 626).

34 Albert Fishlow, "Productivity and Technological Change in the Railroad Sector, 1840–1910," in *Output, Employment, and Productivity in the United States after 1800*, National Bureau of Economic Research, Studies in Income and Wealth no. 30 (New York, 1966), pp. 583–646.

Nevertheless, "Its cumulative character and the lack of a single impressive innovation should not obscure its rapidity. Within the space of some forty years – from 1870 to 1910 – freight car capacity more than trebled. The remarkable feature of the transition was its apparent small cost; capacity increased with only a very modest increase in dead weight, the ratio changing from 1:1 to 2:1. Over the same interval, locomotive force more than doubled as powerful engine types, such as the Mogul, the Consolidation, etc., replaced the familiar and faithful American 4–4–0."[35]

This combination of prosaic, unremarkable improvements, leading to more powerful locomotives and more efficient freight cars, accounted for a reduction of $749 million in operating costs, well over half of the cost savings of $1.3 billion achieved over the period 1870–1910.[36]

Finally, the immense cumulative importance of individual improvements has been pointed to in that most modern of industries, the computer industry. Kenneth Knight, reporting on his own research on the computer industry, asserts that "most of the developments in general-purpose digital computers resulted from small, undetectable improvements, but when they were combined they produced the fantastic advances that have occurred since 1940."[37]

Interindustry relationships

The measurement – even the perception – of the economic payoff to technological innovation is obscured by the difficulties involved in completely identifying the growth in productivity associated with a given innovation. A critical aspect of these difficulties appears to be the preva-

[35] Ibid., p. 635.

[36] Here again we find that these two streams of improvement – in freight car and locomotive – stood in a strongly complementary relationship to each other: "Had the powerful twentieth century engines been developed without that simultaneous remarkable advance in freight-car construction, much more of the increased power would have been dissipated in the nonproductive task of hauling dead weight. A higher ratio of dead weight requires either more or heavier trains to deliver the same payload, both involving additional expense. If 1910 tonnage had to be moved in 1870 freight cars, it would have required about 3.3 of them to equal one 1910 car, and at twice the weight. With identical load factors under both technologies, the same loads would have been carried in four trains of identical weight (but with 3.3 times as many cars) as were actually transported in three" (ibid., p. 641).

[37] Kenneth Knight, "A Descriptive Model of the Intra-Firm Innovation Process," *Journal of Business* 40 (October 1967): 493.

lence, in modern industrial economies, of a special kind of external economy. Specifically, many of the benefits of increased productivity flowing from an innovation are captured in industries *other* than the one in which the innovation was made. As a result, a full accounting of the benefits of innovation must include an examination of interindustry relationships. In part this is due to the fact that industrial development under a dynamic technology leads to wholly new patterns of specialization both by firm and by industry, so that it is impossible to compartmentalize the consequences of technological innovation even within conventional Marshallian industrial boundaries.

One component of these changing patterns of industrial specialization is the emergence of specialized firms and industries that produce no final product at all – only capital goods. In fact, much of the technological change of the past two centuries or so has been generated by these specialist firms.[38] The main beneficiaries of technological change in these capital goods industries are, in the first instance, the buyers of these goods in other industries, but the total benefits may be very widely diffused in an economy of increasingly specialized productive units and high rates of interindustry purchases. The inability to take these interindustry relationships fully into account is a fundamental limitation of most of the recent literature on technological innovation.

It is one of the cardinal merits of input-output analysis that it corrects some of these deficiencies; it breaks open the "black box" in which the primary factors of production, capital, and labor are somehow transformed into a flow of final output and displays a wealth of information on the sectoral flow of intermediate inputs. The technique makes it possible to study the process of technological change by examining changing intermediate input requirements, by looking, that is, "at the coal and ore and steel and chemicals and fibers and aluminum foil; sausage casings, wire products, wood products, wood pulp, electronic components, trucking, and business services that establishments furnish to each other."[39] Many aspects of technological change are visible only at this intermediate level. These take the form of new

[38] See Nathan Rosenberg, "Technological Change in the Machine Tool Industry, 1840–1910," *Journal of Economic History* (December 1963), and "Capital Goods, Technology and Economic Growth," *Oxford Economic Papers* (November 1963), pp. 217–27; and Edwards Ames and Nathan Rosenberg, "The Progressive Division and Specialization of Industries," *Journal of Development Studies* (July 1965), pp. 363–83.

[39] Anne P. Carter, *Structural Change in the American Economy* (Cambridge, Mass., 1970). p. 4.

materials, new machines, new components, or technical processes that never show up in conventional measures of final product for the simple reason that they are not final product. Thus, since highly aggregated approaches jump directly in their reasoning from primary inputs at the beginning of the productive process to final outputs at the end of the process, an enormous amount of interesting information is completely lost from view.[40] The great virtue of input–output analysis, for our present interests, is that it helps us to understand the structural interdependence of the economic system, and the *changes* over time in this structural interdependence, by providing quantitative measures (input–output coefficients) of the interindustry flow of goods and services. Anne Carter has shown that technological change has been associated with an increasing reliance on general sectors – producers of services, communications, energy, transportation, and trade. This has been offset by decreases in other sets of coefficients, most conspicuously in the general, across-the-board declines in the contributions of producers of materials. Technological change has been forcefully associated with a significant expansion in the kinds and qualities of materials and in improvements of design generally. Carter demonstrates how technological change has been expanding the range of substitutability among materials in addition to bringing about an absolute reduction in input requirements per unit of output. The traditional dominance of steel in many uses, for example, has been successfully challenged by aluminum, plywood, and prestressed concrete. The growing importance of plastics and chemicals, and the changes in product design associated with such new and versatile materials, have been clearly quantified. Moreover, technological changes in the crucial area of capital goods, and their increasing complexity and sophistication, have brought a decline in the relative importance of general metalworking and a sharp increase in the role of electrical, electronics, and instrumentation sectors. It is a significant contribution of input–output analysis that such changes can be examined in quantitative terms.

[40] "In earlier days of national income accounting, intermediate production was eliminated to avoid 'double counting.' This is reasonable if one is primarily concerned with measuring an economy's 'success' – the net amount the nation has managed to produce, whatever its methods. Yet this duplicative portion of economic activity is precisely the focus of our present analysis. For it is the composition of interindustry sales that mirrors most directly the effects of changing technology and the organization of production. Intermediate inputs are the specific goods and services used to produce the gross national product. As methods of production change, more of one kind of input will be required and less of another – more chemicals, less steel, and so on – and the interdependence of individual supplying sectors will be changed accordingly" (ibid., p. 33).

In spite of its usefulness, even input–output analysis can capture only a small portion of the kinds of interindustry relationships that are relevant to an examination of the payoff to technological innovation. One would need to include also all the consequences to the customer industry when technological changes in the supplying industry bring about a reduction in the price of the intermediate good. Here, indeed, input–output analysis can at least provide some preliminary guidance on the direction and relative magnitudes of specific innovations. Input–output information enables us to predict that cost-reducing technological changes in some sectors are likely to have wider-range repercussions than similar changes in other sectors. It highlights the pervasiveness of cost reductions in such sectors as transportation, energy, services, and communications, and makes it possible to identify and assess the relative significance of such cost reductions in different sectors of the economy. But the problems are far more subtle and complicated and revolve around the essential fact that technological progress in one sector of the economy has become increasingly dependent upon technological change in other sectors. That is to say, technological problems arising in industry A are eventually solved by bringing to bear technical skills and resources from industry B, C, or D. Thus, industries are increasingly dependent, in achieving a high rate of productivity growth, upon skills and resources external to, and perhaps totally unfamiliar to, themselves.

This situation is not a new one, although it is a phenomenon the relative importance of which has been plainly increasing. It is clearly a function of the growing specialization of industrial activity. In the early stages of the industrial revolution, textile firms produced their own machines. As the size of the market for such machinery grew and as such machines became increasingly complex, the making of textile machinery became the unique responsibility of an increasingly independent set of specialized machinery producers, from whom the textile firms subsequently purchased their equipment.[41] As the textile industry expanded in the nineteenth century, it generated other input requirements that were far beyond its own technical competence and that drew upon the skills of the chemical industry as well as machinery makers. As David Landes has graphically expressed it in connection with the British textiles industry: "The transformation of the textile manufacture,

[41] Nathan Rosenberg, "Technological Change in the Machine Tool Industry," pp. 418–19. The process continues in the twentieth century. Chemical firms used to design their own plants, and still do to some extent. Increasingly, however, they rely on specialized plant contractors for the construction of new plants (see, for example, C. Freeman, "Chemical Process Plant: Innovation and the World Market," *National Institute Economic Review* [August 1968], pp. 30–1).

whose requirements of detergents, bleaches, and mordants were growing at the same pace as output, would have been impossible without a corresponding transformation of chemical technology. There was not enough cheap meadowland or sour milk in all the British Isles to whiten the cloth of Lancashire once the water frame and mule replaced the spinning wheel; and it would have taken undreamed-of quantities of human urine to cut the grease of the raw wool consumed by the mills of the West Riding."[42]

The ways in which technological changes coming from one industry constitute sources of technological progress and productivity growth in other industries defy easy summary or categorization. In some cases the relationships have evolved over a considerable period of time, so that relatively stable relationships have emerged between an industry and its supplier of capital goods. Equipment makers are a major source of technological change in many industries – for example, the aluminum industry.[43] On many occasions the availability of new and superior metals has played a major role in bringing performance and productivity improvements to a wide range of industries – railroads, machine tools, electric-power generation, and jet engines, among others. Since the 1930s the building industry has been the recipient of numerous new plastics products that have found a wide range of uses, not the least of which has been cheap plastic sheeting that made possible an extension of the construction year by providing protection on the building site against inclement weather.[44] The sharp increase in the utilization of commercial fertilizer inputs in American agriculture, so important to the growth of agricultural productivity, can be entirely explained by the

[42] David Landes, *The Unbound Prometheus* (Cambridge, 1969), p. 108.

[43] Merton J. Peck, "Inventions in the Postwar American Aluminum Industry," in *The Rate and Direction of Inventive Activity* (Princeton, N.J., 1962), pp. 279–98, esp. p. 285, table 1.

[44] "In the past thirty years, one new major class of materials has been introduced into the building industry: plastics. Polyvinyl chloride dates from 1936; Polystyrene, from 1938; Melamines, from 1939; Polyethylene, from 1942; Polyesters, from 1952; and Urethanes, from 1953. All of these products have been developed within the chemical industry, many of them as synthetic products for wartime use. The growth of plastics has been rapid. The Census of Manufacturers reports a 1937 volume of $67 million, a 1950 volume of $791.8 million, and a 1958 volume of $1.8 billion . . . De Marco of Monsanto Chemical Company estimated . . . that in 1959 approximately 5 billion pounds of plastic were produced with about 18% going to the construction industry. It is further estimated that the construction industry's consumption rose from 501 to 866 million pounds between 1956 and 1959. About 40% of these plastics were in paints, 20% in laminates and floor coverings, and another 20% in wire coatings and electrical devices and controls" (Little [n. 20 above], pp. 120–1).

decline in fertilizer prices. This decline, in turn, was to a considerable extent the result of technological change in the fertilizer industry.[45]

Often, however, an innovation from outside will not merely reduce the price of the product in the receiving industry but will make possible wholly new or drastically improved products or processes. In such circumstances it becomes extremely difficult even to suggest reasonable measures of the payoffs to the triggering innovation, because such innovations, in effect, open the door for entirely new economic opportunities and become the basis for extensive industrial expansion elsewhere. In the twentieth century the chemical industry exercised a massive effect on textiles through the introduction of an entirely new class of materials – synthetic fibers.[46] Their great popularity, especially in clothing, is attributable to the possibility of introducing specific desirable characteristics into the final product, often as a result of blending (including blending with natural fibers). Thus, materials used in clothing can now be designed for lightness, greater strength, ease of laundering, fast drying, crease retention, and so on.

Technological change in the chemical industry has exercised a similar triggering function in industries other than textiles. In metallurgy, for example, thermochemical and electrothermal developments have considerably widened the range of available metal products by making possible the reduction of the ores of metals with high melting points. The most important instance was, of course, aluminum, but there were others, including manganese, chromium, and tungsten. These latter materials were particularly important in the major materials innovations associated with the development of alloys. In the case of the electrical industry, the chemicals industry played a vital role in the innovation process through the provision of such essential items as refractory materials, insulators, lubricants, coatings, and, with the increasing importance of conductors, through the provision of metals of a high degree of purity. All these profound effects of chemicals innovation have had a relatively limited visibility because of the intermediate good nature of

[45] See Zvi Griliches, "The Demand for Fertilizer: An Economic Interpretation of a Technical Change," *Journal of Farm Economics* (August 1958), pp. 591–606, where a distributed lag model is employed; see also Gian S. Sahota, *Fertilizer in Economic Development: An Economic Analysis* (New York, 1968).

[46] "In the period immediately following World War II there was a major invasion of the textile industry by the chemical industry, as the synthetic fibers and finishes were introduced. From the point of view of technical and economic significance, these have been the major innovations of the last 30 years. In the 50's, then, came a series of innovations involving fabric, yarn, and machinery. Almost all of these (except compacting and tufting) have depended on the chemical innovations of the 40's. The 50's have been a time of minor innovation exploiting the major chemical innovations of the 40's" (Little, p. 56).

most of the products concerned. One could document in detail the manner in which transistors in recent years have been exercising triggering effects similar to the experience of chemicals.

The transmission of technological change from one sector of the economy to another through the sale of intermediate output has important implications for our understanding of the process of productivity growth in an economy. Specifically, a small number of industries may be responsible for generating a vastly disproportionate amount of the total technological change in the economy. Government policy directed at stimulating technological change generally, for example, or for stimulating the output of certain categories of goods or services, will need to be based on the clearest possible understanding of the interindustry relationships that have been discussed in this section. For example, although electric-power generation has one of the very highest rates of technological change and productivity growth of any sector of the economy, the industry has had virtually no R & D expenditures of its own. Rather, technological change in electric-power generation has flowed from the research expenditures of the equipment industry, the metallurgical industries, and various other federally supported research projects. Clearly, any attempt to analyze the economic R&D expenditures must be based on a far better understanding of such relationships than is presently available. For, even though only a few industries are research-intensive, the interindustry flow of new materials, components, and equipment may generate widespread product improvement and cost reduction throughout the economy. This has clearly been the case in the past among a small group of producer-goods industries – machine tools, chemicals, electrical and electronic equipment. Industrial purchasers of such producer goods experienced considerable product and process improvement without necessarily undertaking any research expenditure of their own. Such interindustry flow of technology is one of the most distinctive characteristics of advanced industrial societies.[47] Indeed, it might even be more appropriate to say that such technology flows have radically reshaped industrial boundary lines, and that we still talk of "interindustry" flows because we are working with an outmoded concept of an industry:

> Any consideration of the textile industry would be artificial which did not include the chemical, plastics, and paper industries. Consideration of the machine tool industry must now take into account the aero-

[47] The emergence of the machine-tool industry in the United States in the nineteenth century and the precise role which it played in the development and the interindustry diffusion of the new industrial techniques are examined in Nathan Rosenberg. "Technological Change in the Machine Tool Industry, 1840–1910."

> space, precision casting, forging, and plastics forming industries. These industries are now complex mixtures of companies from a variety of SIC categories, some functioning as suppliers to the traditional industry, some competing with it for end-use functions and markets. "The industry" can no longer be defined as a set of companies who share certain methods of production and product-properties; it must be defined as a set of companies, interconnected as suppliers and market, committed to diverse processes and products, but overlapping in the end-use functions they fill. We can talk about the "shelter" industry and the "materials forming" industry, but we cannot make the assumptions of coherence, similarity and uniformity of view which we could formerly make in speaking of "builders" or "machine tool manufacturers." Similarly, companies are coming to be less devoted to a single family of products and manufacturing methods, and more a diverse conglomerate of manufacturing enterprises, stationed around a central staff and bank, and to some extent overlapping in the markets and functions they serve. These changes are part and parcel of the process of innovation by invasion.[48]

I have emphasized that the benefits of innovation were difficult to identify comprehensively because such benefits were frequently captured by industries other than the one in which the innovation was originally made. The benefits of an innovation may be both highly diffuse and difficult to identify because its availability permits a large number of other alterations (including innovations) in the productive process to take place. Consider the case of electricity. The reduced cost of power alone did not exhaust the productivity benefits of electricity. The social payoff to electricity would have to include not only lower energy and capital costs but also the benefits flowing from the newfound freedom to redesign factories with a far more flexible power source than was previously available under the regime of the steam engine. To appreciate this, it is necessary to cast a much wider net.

Although the rise of electricity began in the 1890s, the industry commenced its rapid growth only after the steam turbine had been brought to a level of efficiency sufficiently high to create the thermal power station and the highly centralized generation of electric power.[49] It was the rise of this new power source that challenged the dominance, at the beginning of the twentieth century, of the coal-using steam engine in industry. The steam engine had always presented serious problems. Its minimum size was too large for small plants. Furthermore, it was highly

[48] Little, p. 181.

[49] Utilities were producing more than half of all electricity in the United States by 1914 (see *Historical Statistics of the United States* [n. 26 above], p. 506).

inefficient in those frequent situations where a large steam engine had to be operated in order to supply small quantities of power. In addition, the steam engine required clumsy belting and shafting techniques for the transmission of power within the plant. These methods were not only responsible for high energy losses; they also imposed serious constraints upon the organization and flow of work, which had to be grouped, according to their power requirements, close to the energy source.

With the advent of the "fractionalized" power made possible by electricity and the electric motor, it now became possible to provide power in very small, less costly units and also in a form that did not require the generation of excess amounts in order to provide small or intermittent "doses" of power. Although the direct, energy-saving and capital-saving effects (including, it should be noted, a vast saving of floor space) were great, the flexibility of the new power source made possible a wholesale reorganization of work arrangements and, in this way, made a wide and pervasive contribution to productivity growth throughout manufacturing. "Shortly after steam power began to yield to electricity, installation of electric motors called attention to the obvious restraints placed on efficiency by the steam engine. Its systems, practices, and factory organization became almost visibly redundant. Thus, as 'unit drive' electric power grew in plant after plant, thoroughgoing reorganization of factory layout and design took place. Machines and tools could now be put anywhere efficiency dictated, not where belts and shafts could most easily reach them. To these advantages were simultaneously added those of revamped industrial processes, leading to mass-production and batch-processing techniques."[50]

Electric power was rapidly adopted by industry. Electric motors accounted for less than 5 percent of total installed horsepower in American manufacturing in 1899. By 1909 their share of manufacturing horsepower was 25 percent, by 1919 55 percent, and by 1929 electric motors accounted for over 80 percent of total installed horsepower in manufacturing.[51] The sharp productivity rise in the American economy in the

[50] Richard B. Du Boff, "The Introduction of Electric Power in American Manufacturing," *Economic History Review* (December 1967), p. 513.

[51] Landsberg and Schurr (n. 25 above), pp. 52–3. They also make the following observations on the efficiency of electric motors: "Installation of electric motors resulted in higher thermal efficiency – a higher yield of mechanical work per unit of primary energy employed in the plant. Instead of less than 10 percent in the case of belt-driven machinery powered by steam engines – or a waste of 90 percent and more between energy input and final utilization – efficiency was 70 to 90 percent in the case of electric motors. With energy losses on the way to the machine practically eliminated, a smaller amount of energy was needed to accomplish the same amount of work (though this was somewhat offset in the economy as a whole by the fact that thermal power generation for its part at then prevailing heat rates, wasted 70 to 80 percent of the fuel's inherent energy)" (pp. 62–3).

years after World War I owed a great deal, directly and indirectly, to the electrification of manufacturing.

But even this account does not exhaust the ways in which electricity contributed to the overall growth in productivity. New patterns of specialization and division of labor became feasible, with important implications for industrial organization. For

> electricity did more than change the techniques and decor of the factory: by making cheap power available outside as well as inside the plant, it reversed the historical forces of a century, gave new life and scope to dispersed home and shop industry, and modified the mode of production. In particular, it made possible a new division of labour between large and small units. Where before the two had almost inevitably been opposed within a given industry – the one using new techniques and thriving, the other clinging to old ways and declining – now a complementarity was possible. Both types could use modern equipment, with the factory concentrating on larger objects or standardized items that lent themselves to capital-intensive techniques, while the shop specialized in labour-intensive processes using light power tools. And often the complementarity became symbiosis: the modern structure of sub-contracting in the manufacture of consumers' durables rests on the technological effectiveness of the small machine shop.[52]

And finally, within the household itself, cheap electricity and small, versatile electric motors were the vital technological breakthroughs that made possible a wide array of household appliances.[53] Although electricity was introduced into the household in its early years almost entirely for purposes of illumination, it soon provided the basis for much else, which eventually transformed the operation of domestic households: refrigerators, electric ranges, water heaters, vacuum cleaners, dishwashers, clothes dryers, freezers, and so on.[54] Indeed, one might therefore well argue that the women's liberation movement is essentially

[52] Landes, p. 288.

[53] The electric-power companies worked very hard at increasing the demand for their output by becoming aggressive salesmen of electricity-using innovations inside the household (see Raymond C. Miller, *Kilowatts at Work* [Detroit, 1957], chap. 11). The value (in millions of current dollars) of electrical household appliances and supplies for selected years was as follows: 1899, 1.9; 1900, 2.4; 1910, 16.3; 1920, 82.8; 1930, 160.0; 1937, 341.0 (*Historical Statistics of the United States*, p. 420).

[54] In an absorbing account of the mechanization of the household, Siegfried Giedion points out that many household appliances – the vacuum cleaner, clothes-washing machine, dishwashing machine – had made their first appearances as early as the 1850s and 1860s. Such devices "belonged to those shelved inventions whose release awaited the coming of the small electric motor" (*Mechanization Takes Command* [Oxford, 1948], p. 553.

due to the combination of declining fertility (in turn partly attributable to a more effective technology of contraception), on the one hand, and the electrification of household chores, on the other. One need not be a technological determinist to argue that the social benefits of the new-found freedom of women in American society are, in large measure, the product of technological innovation.

4 The effects of energy supply characteristics on technology and economic growth

A central feature of industrialization has been the utilization of increasing quantities of energy. A large part of the industrial revolution in Great Britain in the eighteenth and early nineteenth centuries may be adequately summarized as the widespread introduction of techniques for the exploitation of a new energy source – coal – in manufacturing and transportation. The spread of industrialization in the last century or so has had at its core the mastery of a sequence of technologies based upon a succession of fossil fuels – coal, oil, natural gas – and, in the twentieth century, the conversion of fuels and water power into a new energy form, electricity.

As recently as a decade ago, the intimate association between rising energy requirements and economic growth was pointed to as one of those basic facts of industrial life that had to be understood and accepted by all industrializing countries. But as yet, no minatory tone was used in making the historical association. Increasing energy use was simply one of the correlates of industrialization – like higher rates of capital accumulation, increasing education, urbanization, and declining infant mortality – that had to be appreciated and planned for if a country was to achieve economic growth.

The decade of the 1970s has, of course, changed all that. The impact of the Arab oil embargo and the subsequent emergence of OPEC as the most powerful cartel in history have focused attention as never before upon the question of energy supply. Like the good Dr. Johnson's observation about an approaching appointment with the hangman, OPEC has served wonderfully to concentrate the mind – in this case, upon the adequacy of energy supplies to sustain continued economic growth.

If there was an air of casual complacency about energy supplies at the beginning of the 1970s, it was thoroughly dispelled within a few years

This paper was prepared for and presented at the Electric Power Research Institute workshop on Energy, Productivity and Economic Growth, Palo Alto, California, January 12–14, 1981. It is included in the published proceedings of that conference (forthcoming).

by the discovery of limits. Although several such limits were discovered (or perhaps, in deference to the shade of Malthus, we should instead say "rediscovered"), energy supplies became the most imminent threat to growth prospects.

As a result of this intense preoccupation with energy supplies, there have been numerous attempts to consider more systematically the relationship between energy and economic growth. Obviously, we are dealing with magnitudes that cry out for quantitative expression. Energy models that can be taken as serious guidelines to alternative futures cannot be constructed without such quantification. At the same time, such energy-modeling exercises are likely to be full of pitfalls. This is especially so when such modeling takes place at a high level of aggregation and when different sources of energy are aggregated by a reduction to some common denominator such as British thermal units (Btus).

The purpose of this paper, therefore, is to examine several aspects of the past relationship between energy supplies and economic growth. A key feature of the following discussion is the critical role of technical change in mediating the relationship between energy supplies, on the one hand, and the changing needs of industrializing societies, on the other. If future policies are shaped by energy models that seriously misspecify the relationship between energy and growth, or that focus narrowly upon some engineering criterion, such as minimizing Btus by selecting a specific energy alternative, the results are likely to be elaborate exercises in suboptimization. Thermodynamic efficiency, which may make a great deal of sense in designing an engine, may make no sense at all when applied to an economic system or its subcomponents. Thus, the implications of this paper will make the construction of credible energy models more difficult. But that consequence, I believe, is a reflection of certain properties of the real world – and of the operation of a dynamic technological system in particular.

Because of time and space limitations, this paper will consist of two main substantive portions: first, a discussion of technical change and energy use in the metallurgical sector and, second, a discussion of electric power and its utilization. The justification for these choices of emphasis is simple. The metallurgical sector (Primary Metal Industries, SIC 33) was for many years the largest single industrial user of energy and is now the largest user of electricity. Electric power has become a dominant and preferred energy form in many sectors of the economy. Furthermore, there is a partial convergence of the two sectors in the late nineteenth and twentieth centuries because electricity has come to play a uniquely important role in certain portions of the metallurgical sector. Indeed, the very fact of that convergence provides important support for a central argument of this paper. In both cases, as I will suggest, a preoccupation with narrow conceptions of efficiency or with static models of energy use patterns will lead to choices or policies that are socially suboptimal. Indeed, I perceive that the major consequence of an increasingly strident advocacy of greater energy efficiency, narrowly conceived, will be to institutionalize policies of suboptimization.

Metallurgy

The evolution of the metallurgical industries over the past two centuries or more can be grasped only very imperfectly by exclusive concentration upon the cost of alternative fuel sources. Fuels have never been perfect substitutes for one another. Rather, different fuels have displayed special, sometimes unique, characteristics, and the search for special characteristics, or combinations of them, has decisively influenced both the search for new technologies and the selection from among the technologies available at any particular time. When correctly viewed, this is *not* a distinction between attempting to minimize production costs and pursuing other kinds of economically irrational goals. Rather, the essential point is that these other characteristics were of great economic significance; as a result, the fuel source that was cheapest in Btu terms was often not identical with the technological alternative that minimized the final costs of metal production.

A distinctive aspect of the metallurgical industries is that their productive processes are highly fuel-intensive. Metal ores appear in nature in various chemical combinations with less desired materials. The processes by which they are separated and refined have been the object of untold human exploration and ingenuity – and frustration. A decisive consideration in the emergence of particular metals has been the ability to develop technologies that could separate them without incurring outrageously high fuel costs. As a consequence, the industrial use of certain metals has not simply reflected their abundance in nature. There is, for example, far more aluminum in the earth's crust than iron. Nevertheless, iron has for many centuries been the most commonly used of all metals, whereas aluminum was only a laboratory curiosity in the nineteenth century and has been a significant industrial metal for less than a century. The reason is that iron can be smelted in blast furnaces and extracted at reasonable costs in terms of fuel and labor. Aluminum ore, in contrast, cannot be smelted in a blast furnace, even though it is similar chemically to iron ore. The only commercially practical ways of extracting aluminum involve huge quantities of energy. Moreover – and this is essential to the analysis of this paper – that energy must be delivered *in a particular form.* I will return to this point later.

A major reason, then, why iron and, later, steel came to be the predominant metals of industrializing societies is that iron lent itself to a smelting process that a more abundant metal, aluminum, did not. This smelting process had the essential characteristic of producing iron by using acceptable quantities of fuel. Even so, the fuel costs of iron production were so great before the industrial revolution that iron was employed very sparingly.

A central aspect of the industrial revolution was the development of new metallurgical technologies that were far more fuel-efficient than earlier technologies had been. However, to reduce these innovations merely to greater fuel efficiency is to exclude the most essential part of the transformation. The point is that the increasingly efficient use of fuel came only with a shift from an old fuel source, wood (charcoal), to a new one, coal (coke).

Thus, given the energy-intensive nature of metallurgy, a major theme in its technical history has been the search for greater efficiency in its utilization of fuel. But that search, in turn, has involved not just (1) the discovery of ways of reducing fuel requirements per unit of iron output from existing fuels, although that has been an important part of the story, but also (2) the development of techniques that would make it possible to exploit *more abundant* fuel sources. An additional, and closely connected point, is that (3) technical change in iron production has involved a continual search for more effective means of controlling the quality of the final product. This was always a serious problem, because the central productive processes in the iron industry involve subjecting nonhomogeneous materials to intense heat in order to achieve chemical transformations that were very inadequately understood. Yet the usefulness of the final product and its performance characteristics were powerfully affected by even minute variations in the chemical composition of the iron ore or the fuel. Finally, (4) technical progress involved a search for fuels that possessed other characteristics – physical or chemical – that were essential to the success of a new technology and the teasing out of further productivity improvements. Each of these features is important in understanding the relationship between energy inputs and output of metallurgical products.

Consider the substitution of coke for charcoal in eighteenth- and nineteenth-century blast furnaces. A main reason for the highly decentralized location of the early iron industry was that it was dependent upon a fuel – wood – that it consumed in large quantities and that was itself widely distributed geographically.[1] But there were other considerations at work. Charcoal disintegrated rapidly when it was transported. Therefore, apart from the high transportation cost, the physical characteristics of charcoal dictated that it be utilized reasonably close to the ovens.

In addition, so long as the blast furnace fuel was charcoal, it was impossible to take advantage of the most important direction of improvement in blast furnace technology – making the blast furnace taller. The reason again pertained to a simple characteristic of charcoal as a fuel: It was readily crushed and could not support a heavy charge. Thus, charcoal blast furnaces had to be small. As one authority has stated, speaking of the iron-smelting technology of the late Middle Ages: "Wood charcoal is neither strong, tough nor hard; it is friable and is likely to break down into powder when subjected to even fairly light loads. It is unable therefore to support the weight of the burden in a very tall furnace, a fact which effectively limited the practicable height of the Stückofen. The limitation imposed by this factor persisted for a further three hundred years."[2]

The severity of the constraint upon blast furnace output resulting from limitations upon stack size is obvious from estimates of annual output. The average annual output of British blast furnaces in the early eighteenth century was about 300 tons. There were various reasons for this small output, but the most important was the size limitation imposed by the nature of the charcoal fuel.[3]

With the introduction of coke, and its capacity for supporting a much heavier charge than charcoal, it was possible for blast furnaces to be enlarged to greater heights. The problem then became the need for more powerful blasts to sustain combustion. This problem was eventually solved by the introduction of the steam engine, together with blowing cylinders, to create the stronger blasts required by the larger furnaces.

[1] The location of the blast furnace before the arrival of the steam engine was also influenced by the desirability of placing where it was possible to harness water power to drive the bellows. Even the very modest height that blast furnaces attained before the industrial revolution would have been impossible without the higher blast pressures made possible by water-driven blowing engines. See Leslie Aitchison, *A History of Metals* (MacDonald and Evans, London, 1960), vol. II, pp. 342–3. As Aitchison points out, "the successful use of the heightened *Stückofen,* and the production of molten iron, depended entirely upon the higher blast-pressures that could be attained by using water-driven bellows. In the absence of such pressures the taller furnaces would have clogged up because the temperatures attainable within the upper parts of the shaft would have been insufficiently high." Ibid., p. 343.

[2] Ibid., p. 343.

[3] Charles Hyde, *Technological Change and the British Iron Industry, 1700–1870* (Princeton University Press, Princeton, N.J., 1977), p. 9. Hyde states that the maximum practical height of a blast furnace was twenty-five feet. For various reasons, the blast furnaces were operated for only about thirty weeks per year.

Thus, the introduction of coke into the blast furnace in the eighteenth century liberated metallurgy from dependence upon an increasingly expensive fuel and substituted a far more abundant one. In addition, the physical characteristics of the new fuel – its greater hardness and strength – provided the basis for a transformation in metallurgical technology that was to extend over two centuries: the growth in furnace size and the related improvements in fuel efficiency that were dependent upon that growth.[4] These developments could not have occurred in a charcoal regime, however abundant the charcoal.

The introduction of the new fuel brought with it a host of new problems. Coal is a highly heterogeneous mineral, and even small variations may have highly significant consequences. In fact, coal as a commercial fuel took so long to introduce because the substitution of coal for wood or charcoal often had highly deleterious consequences on the final product – in brewing, baking, glassmaking, and the metallurgical industries generally.[5] The introduction of coking in the seventeenth century helped a great deal, as did the design of furnaces (such as crucible or reverberatory) that prevented direct contact between the fuel and the product. Nevertheless, specific properties or chemical components in the coal, especially the sulfur content, continued to plague metallurgists. In many end uses, iron made from charcoal continued to be specified by the buyer right through the nineteenth and into the twentieth century. Indeed, two charcoal blast furnaces were actually constructed in Allentown, Pennsylvania, as late as 1910.[6] Even very small differences in coal composition continue to be very important for metallurgical purposes. The Chicago Board of Trade still does not deal in long-term coal contracts because of the difficulty of acceptable, standardized grading of metallurgical coal. Given the large number of exacting specifications written into these contracts, a futures market for such coal remains impossible.[7]

[4] "Coke is obviously harder and stronger than charcoal, but it is also harder and stronger than the coal from which it springs and is therefore much less prone than coal to disintegrate within a heavily burdened furnace and so to choke it." Ibid., p. 445.

[5] John Nef, "The Progress of Technology and the Growth of Large-scale Industry in Great Britain, 1540–1640," *Economic History Review*, 1934, pp. 3–24.

[6] Victor Clark, *History of Manufactures in the United States* (McGraw-Hill, New York, 1929), vol. III, p. 72.

[7] Mr. Benhardt Wruble, executive director of the President's Interagency Coal Export Task Force, was recently quoted as saying of metallurgical coal ordered for export, "People order it like fine wine or tobacco, such as Vein 3, Ace Mine." *New York Times*, October 26, 1980, p. 24. The prospects for a futures market in steam coal, which is not so precisely specified, appear somewhat more favorable.

The shift from a wood-based iron industry to one based upon coal was virtually complete in Great Britain by 1800. The introduction of coke into the blast furnace, which had begun early in eighteenth century but progressed only very slowly before midcentury, was complemented by Henry Cort's puddling process and rolling mills in the 1780s.[8] The puddling process did more than substitute energy for labor. Since its central feature was the use of a reverberatory furnace that separated iron from fuel, puddling could tolerate the use of more impure fuels. As a result, it became possible to refine iron by using a new fuel – coal – that was highly impure chemically by comparison with charcoal but also available in far greater supply.[9]

The puddling process possessed another vital characteristic: It could employ a wider range of iron inputs than the other techniques then employed to make wrought iron. Specifically, it could make use of scrap iron. The ability to reuse scrap is an important fuel-saving characteristic of a technology and one that would play a particularly significant role in later metallurgical history. At the same time, the ability of a technology to exploit a wider range of inputs than earlier technologies has a fundamental economic significance. It means that the new technology is effectively enlarging the resource base itself.

This history of the iron and steel industry since 1800 involves a complex interplay of several forces, an important one being the search for greater fuel efficiency.[10] Two things can be said: First, that search was so successful that the relation between fuel requirements and output was never stable for long. At any time, an extrapolation two or three decades into the future, based upon the assumption of reasonably stable input–output coefficients, would have been seriously off-target. Second, an understanding of what was shaping the structure and performance of the industry could never be adequately grasped by concentrating *solely* upon the relationship between aggregate fuel requirements

[8] For a careful recent reevaluation of the evidence on technological change in the British iron industry, with particular attention to the timing of the major innovations, see Charles Hyde, *Technological Change and the British Iron Industry, 1700–1870* (Princeton University Press, Princeton, N.J., 1977).

[9] The quality of the fuel was much more important in the blast furnace, where the fuel and the iron ore came into direct contact and where fuel impurities were passed directly into the final product as a result.

[10] It is important to remember that not all improvements have required new hardware. The eventual development of the integrated steel mill was an organizational innovation the central feature of which was the relocation of the successive stages in steelmaking so as to eliminate the huge heat loss previously imposed by the necessity of reheating and the failure to utilize waste gases. The resulting fuel savings from bringing together coke ovens, blast furnaces, and open hearth furnaces were very large.

and aggregate final output. Qualitative considerations regarding both inputs and outputs, the unavoidable concern with specific physical and chemical characteristics, and technological innovations that redefined the significance of the resource environment for the industry – all these are not minor themes but essential parts of the story.

Consider the impact of the hot blast, introduced into blast furnace practice in Britain by Neilson in 1828. The hot blast was an extremely simple invention; yet, because of the enormous savings in fuel that it made possible, it was one of the most important technical improvements in the iron industry in the nineteenth century. Preheating the blast before entering the furnace saved fuel.[11] When the waste gases from the top of the furnace were employed in heating of the blast, the fuel economy was much greater. When the hot blast was introduced into the United States during the 1830s, it became apparent that its benefits were not confined to saving conventional blast furnace fuel. The hot blast also made it possible to utilize anthracite coal. Since large supplies of this coal were located near the major population centers east of the Appalachians, the introduction of the hot blast was an event of great economic importance.[12] For it made possible the exploitation of anthracite for metallurgical purposes and thus greatly expanded the resource base of the iron industry.[13]

The hot blast achieved economic importance in America partly because it was able to employ a form of coal that had previously been unusable for smelting. Anthracite is not only very hard but also has a low gas content. This low gas content, in turn, made combustion of the coal very difficult until the advent of the hot blast. Within fifteen years

[11] "The use of the hot blast raised the temperature of the furnace, brought about a more complete combustion of the fuel, and lowered fuel consumption. Higher temperatures also speeded up the smelting process, making possible much higher output from a given furnace" Hyde, *Technological Change,* p. 147. The hot blast also made possible the smelting of lower-grade ores.

[12] Alfred Chandler has in fact argued that America's industrial "takeoff" in the 1840s was closely linked to the exploitation of the anthracite mines of eastern Pennsylvania. See Alfred D. Chandler, Jr., "Anthracite Coal and the Beginnings of the Industrial Revolution in the United States," *Business History Review,* Summer 1972.

[13] In Britain the impact of the hot blast was an inverse function of coal quality, where quality is measured by carbon content. The innovation offered a greater cost saving to regions with poorer (i.e., lower carbon content) coal. Thus, the hot blast offered great savings in fuel costs to Scotland, where the quality of coal was poor, and small savings to South Wales, which had the best-quality coal in Britain. As Hyde shows. Scottish ironmasters were the most rapid adopters of the innovation, and South Wales ironmasters the slowest. Hyde *Technological Change,* chapter 9.

after its introduction, more than 45 percent of pig iron production in the United States was accounted for by anthracite.[14]

Although the growth of the American iron industry thus drew very heavily upon anthracite in the antebellum years, its use as a blast furnace fuel declined after the Civil War.[15] Anthracite could not compete successfully with good coking coal, such as had been discovered in the Connellsville region of western Pennsylvania. The reasons, again, have to do with the physical characteristics of the coal that render it distinctly inferior, at least inside a blast furnace, to coke.

> Anthracite was a dense fuel, and although the lack of bituminous gas is common to anthracite and coke, the often-heard characterization of anthracite as a natural coke is not exact. For the lack of gas makes anthracite a very dense material, and the absence of gas makes coke a very porous one. When bituminous coal is coked, its volume does not fall. Rather than collapse upon itself, the remaining material in a good coking coal forms a fine honeycombed structure with air spaces in between. This means that coke has a higher surface area in relation to its volume than anthracite, and that it can burn faster. If a greater volume of air is blown into a blast furnace, coke will burn faster, while anthracite will speedily reach a limit set by the maximum rate of combustion on its restricted surface area.[16]

Thus, the chemical "purity" of anthracite, which has often caused it to be regarded as a high-grade coal, made it a very limited and inadequate source of fuel. The physical structure of the coal rendered it incapable of achieving rapid combustion. Yet, rapid combustion was critical to a major source of productivity improvement after 1870, that of "hard driving." In the hard-driving method, the output of a blast furnace is increased by operating it under higher pressure. For such purposes, anthracite, however attractive its chemical composition, was unsatisfactory. What was needed was a coal that could resist crushing but was capable of rapid combustion.[17]

Shortly after the mid-nineteenth century, several major innovations created the transition from iron to steel. The Bessemer process was the first and most spectacular – certainly in a visual sense; the effect of blowing air rapidly through liquid pig iron left an enduring impression on all who saw it. Bessemer's process greatly reduced the fuel cost of refining, partly by taking advantage of the impurities in the pig iron by

[14] Peter Temin, *Iron and Steel in 19th Century America* (MIT, Cambridge, Mass., 1964), p. 52.
[15] Ibid., Table C.3, pp. 268–9.
[16] Ibid., p. 201. [17] Ibid., pp. 157–63.

using them as fuel.[18] Indeed, Bessemer's 1856 paper presented to the British Association for the Advancement of Science, in which he announced his new method, bore the intriguing title "Manufacture of Malleable Iron and Steel without Fuel."

Blowing air through the molten iron removed two of the impurities – carbon and silicon. Unfortunately, it did not remove phosphorus, and the Bessemer process was confined for many years to the use of nonphosphoric ores. Indeed, the widespread use of the Bessemer process dramatized the problems of quality control in metallurgy with respect to both inputs and outputs. Not only did this method of steelmaking require phosphorus-free ores, but Bessemer steel tended to become brittle as it aged. This deterioration in quality had much to do with the declining popularity of Bessemer as other methods became available and as new products were introduced that required high-quality steel. Although nineteenth-century metallurgists never established the cause, it turned out that blowing air through the molten iron not only removed impurities but *added* one of its own. The brittleness was caused by minute quantities of nitrogen from the air that dissolved into the iron during the blowing.

The Bessemer process had drastically reduced refining costs and thoroughly dominated American steelmaking in the last two decades of the nineteenth century; around 1880, almost 90 percent of all steel manufactured in the United States was Bessemer steel. By the time of the First World War, this proportion had declined to one-third.[19] The Bessemer process suffered from difficulties in quality control due to the narrow chemical specification of inputs required for best results (nor could it use scrap), a method of production that added a significant impurity while removing others, and, oddly, the great speed of the combustion process, which made quality control even more difficult. Thus, to focus exclusively on the energy efficiency of the Bessemer process and to ignore these other characteristics is to overlook some of the most critical advantages and disadvantages of this technology compared to those that eventually displaced it.

[18] Writing on the use of the Bessemer process in the American rail industry, Temin points out: "It took 7 tons of coal to make 1 ton of cast (crucible) steel from pig iron by the old method: 2½ tons of coal to make blister steel and 2½ tons of *coke* to make cast steel. The same process with a Bessemer converter required ⅟₁₀ of a ton of coal to heat the converter plus about %₁₀ of a ton to melt the iron. One part fuel in the Bessemer process equaled 6 or 7 parts in the old method of steelmaking, and a comparison of labor and machinery requirements would yield similar results." Ibid., p. 131 (emphasis in the original).

[19] Ibid., p. 4. Much of the decline was associated with the decreasing demand for rails with the completion of the country's railway network. Around 1880, 80 percent of Bessemer steel was devoted to the manufacture of rails.

The major post-Bessemer innovations in iron and steel continued to emphasize not only greater fuel efficiency but also continual widening of the range of usable inputs and more precise quality control in the final output. The open-hearth furnace was essentially an extension of the earlier puddling furnace. It provided higher heat levels and achieved greater fuel efficiency than its predecessor by utilizing previously unexploited heat emitted by waste gases. "The open-hearth furnace . . . was a puddling furnace to which greater heat could be supplied and, consequently, in which there was no need for a puddler."[20] As compared to the Bessemer process, which worked most effectively within a relatively narrow range of ores, the open-hearth process could exploit a much wider range of inputs, including increasingly abundant scrap as well as a far larger range of the immense iron ore deposits around Lake Superior. In this respect, its superiority to Bessemer cannot be adequately conveyed by static cost comparisons. Furthermore, the introduction of the Gilchrist–Thomas "basic" process in 1879, which was applicable to both the Bessemer and open-hearth processes, further tilted the balance in favor of the open hearth. The essential point here is that the basic process was far more liberating to the open-hearth than to the Bessemer process, since ores suitable to basic Bessemer exist in only very modest amounts in the United States. By contrast, the basic process opened up enormous reserves of phosphoric ores for which the open-hearth process had been previously unsuitable.[21] Finally, the open-hearth process was far superior to Bessemer in the critical matter of quality control. This was due partly to the much slower pace of the process.[22] It could produce steel to the increasingly exacting specifications required for many of the new machines and products of the twentieth century.

The basic oxygen furnace, which was first introduced in the mid-1950s and now constitutes the major share of U.S. steelmaking, has achieved a

[20] Ibid., p. 139.

[21] Ibid., pp. 138–45. In Europe the basic process was responsible for a great shift in favor of the continent vis-à-vis Britain. Britain had few deposits that were well suited to the innovation, whereas Germany and Belgium had huge supplies of suitable phosphoric ores.

[22] "The open-hearth process permits closer metallurgical manipulation and control than is possible in the Bessemer process. During the open-hearth process frequent samples are taken for chemical analysis and other tests, and extensive use of instruments and scientific apparatus permits regulation of temperature, furnace atmosphere and conditions." W. N. Peach and James A. Constantin, *Zimmermann's World Resources and Industries*, 3rd ed. Harper & Row, New York, 1972), p. 455. The action of the Bessemer converter might be completed in nine to fifteen minutes, whereas the open hearth process lasted for about ten hours. Here again, apparently secondary characteristics of the manufacturing process turn out to have vital consequences.

drastic reduction from nine to ten hours of the open hearth to forty-five minutes. Quality control is maintained and metallurgical transformations are monitored, in spite of the great increase in speed, through elaborate techniques of instrumentation. "A computer provides the operator with complete information on the weights of hot metal, scrap and flux, as well as the volume of oxygen needed to achieve given specifications in the end products."[23] The basic oxygen furnace reduced both capital and operating costs and became feasible only when technological innovation made it possible to produce oxygen cheaply and in very large quantities. It had long been understood, even by Bessemer, that the use of oxygen could bring great benefits to steelmaking, but the cost of oxygen remained prohibitively high until after the Second World War.

The subtle interaction between energy requirements and qualitative factors in the steel industry, and the indispensability of examining energy developments within a larger systematic framework, have become clearly apparent in the last few decades as a result of the slow decline in the quality of available Lake Superior iron ores.[24] It had been widely assumed that rising energy costs were unavoidable as the steel industry moved down the quality gradient, and that rising costs of its energy-intensive final product were also inescapable. However, technical changes involving the pelletization of low-grade taconite ores (about 25 percent iron content) have made possible substantial energy (and labor) savings per ton of molten iron at the blast furnace compared to the older, more conventional techniques exploiting the naturally concentrated hematite and goethite ores (50 percent or more iron content). In fact, between 1954 and 1975 the shift toward pelletization was responsible for a net energy savings of 17 percent.[25] This net saving has been accomplished in spite of the fact that the pelletization process is highly energy-intensive. Indeed, in the mining stage alone, where pelletization is carried out, *six times* as much energy is consumed compared to extraction of the naturally concentrated iron ore.[26] The point is, of

[23] William T. Hogan, S.J., *The 1970s: Critical Years for Steel* (Lexington Books, Lexington, Mass., 1972), p. 40.

[24] For an interesting discussion of the stagnant productivity in the American blast furnace in the mid-nineteenth century, before the exploitation of the Lake Superior ores, see Robert C. Allen, "The Peculiar Productivity of American Blast Furnaces, 1840–1913," *Journal of Economic History* September 1977. Allen argues that the productivity gap that occurred from 1840 to 1870 between American and European blast furnace performances was due to the siliceous nature of American ores. The gap rapidly disappeared with the introduction of Lake Superior ores.

[25] Peter Kakela, "Iron Ore: Energy, Labor, and Capital Changes with Technology," *Science,* December 15, 1978, p. 1151.

[26] Ibid., p. 1152.

course, that this larger energy expenditure in one part of the steel industry was more than offset by resulting economies elsewhere. As a result, an energy "accounting" that looked only at one sector rather than at the entire system would be extremely misleading. Pelletization was responsible for reduced transportation costs since less waste rock was being moved compared with naturally concentrated ores. Far more important, however, was the fact that pelletization was responsible for great energy savings at the blast furnace. This was due to the increased chemical efficiency of the blast furnace as a result of improved permeability.[27]

Thus, when *all* energy requirements of steel production are taken into account, energy costs per ton of steel output have declined in the post-World War II years in spite of the substantial decline in the quality of iron ore. But, from a larger perspective, technological change in the postwar years has considerably enlarged the resource base over that of 1945 or 1950 by developing commercially feasible methods of utilizing resources once considered submarginal. As a result, the usable iron ore resources of the United States are, by any reasonable economic criterion, larger than they were thirty years ago.[28]

Electricity

Thus far, we have looked at a single industrial sector, iron and steel, in order to explore the historical connections between energy and economic growth. I propose now to alter the focus and to concentrate upon a single energy *form*, electricity. In spite of the shift in focus, some common elements persist. Expressed in most general terms, the history

[27] "The substitution of pelletized ore for naturally concentrated iron ore causes three main energy changes in blast furnace firing: (i) the gas/solid contact ratio is increased because of improved permeability obtained with pellets, (ii) more air can be blown into the blast furnace because of the improved permeability, and (iii) waste removal (slag) is reduced because of the richer content of the pellets. Each of these changes reduces the amount of coke required per net ton of molten iron." Ibid, p. 1153.

[28] The instructive story of the prolonged research enterprise leading up to the eventual commercial success of pelletization of taconite ore in the early 1960s is told in E. W. Davis, *Pioneering with Taconite* (Minnesota Historical Society, St. Paul, 1964). Even though no new fundamental knowledge was required, the solution of innumerable engineering and processing problems required many years of tedious experimentation. As Davis points out: "The processing of taconite is simple in theory but complex in execution. The extremely hard rock must be crushed and ground to a fineness resembling flour. This fine grinding liberates the small particles of high-grade magnetite. These are caught and removed (or concentrated) by magnetic separation. The particles must then be put back together (agglomerated or pelletized) to make pieces large and hard enough for shipping and smelting." Ibid., p. 69.

of electricity reveals a continuous search, now on the part of many industries, not just for the cheapest source of energy but for a form of energy possessing certain characteristics that are economically desirable. The form in which the energy is available can be translated into tangible cost reductions. Sometimes these cost reductions are of total energy requirements, in parallel with the pelletization of taconite, in which the introduction of energy-intensive techniques in one place (mining) generated more than compensating savings elsewhere (in transportation and in the blast furnace). In other cases, electricity may offer superior opportunities for control and manipulation of heat or mechanical energy that other fuels cannot match.[29]

In discussing electricity as a form of energy, the purpose is to facilitate comparison of the merits, and the possibilities for substitution, among a variety of energy forms. It is therefore important to observe that, for a wide range of technologies, there is no alternative energy form, at least not at the present state of knowledge. All existing forms of fast, long-distance communication – the telegraph, telephone, radio – involve the transmission of electrical impulses. Although the quantities of electricity involved are small compared to usage by other industries, these are industries that are intrinsically electricity-based – as are many others. Historically, the great initial impulse to the growing use of electricity and the emergence of the central power station was the invention of the incandescent lamp. Although one can now point to substitutes – such as the gas lamp, which the electric lamp displaced – such lighting is rooted in a particular energy form and could hardly exist without it. There are, of

[29] This superior ability to control heat seems to be the distinct advantage of a new electric melting furnace that Corning Glass has recently introduced: "The Corning Glass Works . . . the country's largest producer of specialty glasses and a leader within the industry in energy conservation techniques, has developed a melting furnace that company officials maintain is three times as thermally efficient as its gas-fired counterparts.

"In conventional glassmaking, the loose, dry 'batch' material enters a horizontal tank-like furnace from the rear, and as it moves forward, is turned molten by blasts of natural-gas-fueled fire criss-crossing the top of the molten material. This can be very inefficient, because much of the heat escapes from the top of the furnace, never affecting the batch.

"The Corning furnace, on the other hand, is a vertical cylinder with electrodes embedded in the sides. An electric charge flows through the molten glass, from electrode to electrode, concentrating heat in the center of the furnace. And since the raw material is poured in from the top, the exposed upper surface of the batch is always cool, forming a sort of insulation that keeps heat from escaping through the top of the furnace.

"Officials of Corning, which operates a total of 130 furnaces, noted that the company had reduced its dependence on natural gas from 85 percent to 70 percent of production using this method, while cutting its energy costs by one-third, or around $25 million." *The New York Times*, November 27, 1980, p. 24.

course, various ways, and a range of substitute fuels, for the generation of electricity, but it is a basic fact that in our present industrial civilization numerous technologies depend upon electricity and that no close substitutes for many of these technologies presently exist.[30]

In metallurgy, such is not the case. As we have seen, metallurgy became a primarily coal-based technology in the nineteenth century. Yet, one of the most conspicuous aspects of twentieth-century metallurgy has been the growing reliance, in specific areas, upon electricity as an energy source.[31] The electric furnace was first developed in the late nineteenth century but it was long confined to a limited number of specialty steels in which the furnaces produced only a few tons per heat. Again, the distinctive advantage was one of quality control. Some contamination is unavoidable whenever fuel is burned in a blast furnace. The electric furnace offered freedom from such sources of contamination that was essential to the production of high-quality alloys. As a result, this technique occupied a small but highly significant niche in the industry as the optimal way of producing a variety of alloy steels.

After World War II, the electric furnace was employed in larger sizes, predominantly to produce carbon steel. Capital costs are much smaller than those for other steelmaking technologies. Of great significance is the flexibility of the furnace with respect to inputs. It can even operate with 100 percent scrap charge and can be established independently of blast furnaces and coke ovens. Thus, the electric furnace is now a highly attractive way of making relatively inexpensive additions to steelmaking capacity.[32] Moreover, it provides greater locational flexibility than existed before. It can be located far from coalfields or ore deposits, wherever large "deposits" of urban junk are available for mining.[33] The electric furnace is therefore a highly decentralizing technology.

[30] In addition, the use of electricity-intensive processes is sometimes instrumental in providing striking improvements in overall energy consumption. One of the most dramatic examples is the production of enriched uranium by the electricity-intensive method of gaseous diffusion. A single ton of enriched uranium requires 1 million kilowatt-hours to produce. But that ton of enriched uranium, used in a nuclear plant to produce electricity, will produce as many as 32 million kilowatt-hours of electricity power. See Saunders Miller, *The Economics of Nuclear Power* (Praeger, New York, 1976), p. 10.

[31] The same might also be said of the chemical industry, where electrolytic techniques have become so important. Electrolysis is also employed in metallurgy, as in the refining of copper.

[32] Hogan, *The 1970s*, p. 41.

[33] The recycling of scrap is now a major factor for the steel industry as a whole. From January to July 1979, "total purchased scrap provided about 53% of the total metallic charge (pig iron and scrap) in the production of iron and steel." U.S. Department of Commerce, Industry and Trade Administration, *1980 U.S. Industrial Outlook for 200 Industries with Projections for 1984* (Washington, D.C., 1980), p. 170.

Thus, the electric furnace offers the unique opportunity of bypassing the highly energy-intensive earlier stages of mining, coke making, and smelting. Where scrap is available, the electric furnace is a highly energy-saving technology. Such furnaces were used to process 36 percent of all scrap iron and steel in 1977.[34] Accordingly, it is not surprising that some of the largest integrated steel producers, such as the Bethlehem Steel Corporation and the Jones and Laughlin Steel Corporation, have recently expanded their capacity by the addition of electric furnaces.

The electric furnace, which constituted less than 20 percent of the steelmaking capacity in 1970, accounted for 22 percent of total steel production in 1977 and is currently responsible for more than one-quarter of all steel production in the United States. Steel production by the electric furnace has exceeded open-hearth production since 1977.[35] The increasing attractiveness of the electric furnace is based upon one further characteristic that has so far been ignored but has become increasingly significant in recent years: its relatively modest polluting effects in an industry notorious for pollution in the past. The increasingly tight emission control standards of federal, state, and local governments have become a prime economic factor in the continued viability of the industry, and the air and water pollution of the electric furnace are far easier to deal with than that of the older technology.

The specifically electrical form of energy makes another very different, and easily overlooked, contribution to the high levels of productivity in the steel industry. Most steel today is produced in the basic oxygen furnace. This method, as was mentioned earlier, became commercially feasible only with the availability of relatively cheap oxygen. The technology for producing oxygen – by liquefaction and rectification of air – is highly electricity-intensive. A typical plant requires 330 kilowatt-hours of electricity for each ton of oxygen produced.[36] Electrically driven condensers and pumps are essential to modern oxygen production technology. Seventy percent of the output of oxygen is used in iron and steel manufacturing.[37] Thus almost all of the output of the steel industry today is critically dependent upon electricity – directly in the case of the electric furnace and indirectly in the dependence of the

[34] U.S. Department of Interior, Bureau of Mines, *Minerals Yearbook 1977*, Vol. 1 (USGPO, Washington, D.C., 1980), p. 537.

[35] Ibid., p. 519; *Wall Street Journal*, December 29, 1980, p. 14.

[36] R. Norris Shreve and J. A. Brink, *Chemical Process Industries* (McGraw-Hill, New York, 1977), p. 110.

[37] U.S. Department of Interior, Bureau of Mines, *Mineral Facts and Problems* (USGPO, Washington, D.C., 1975), p. 763

dominant technology, the basic oxygen furnace, upon large quantities of oxygen produced by an electricity-based technology.

Aluminum, now the second most important primary metal in the American economy, demonstrates forcefully the dominant importance of specific characteristics of different energy forms. It is a commercial product that was made possible by electricity. Although electricity costs have been greatly reduced within the past decade,[38] the attempt to substitute other energy forms has so far been a total failure, and current research offers no short-term prospects for success.[39] The aluminum industry's inability to find a satisfactory substitute for electricity seems every bit as confirmed as the steel industry's inability to find an acceptable substitute for coal in the blast furnace – although research in these areas has been conducted for decades.[40]

Aluminum is a relatively new industry. The element was first isolated by Hans Christian Oersted in 1825, but it remained little more than a curiosity for a long time. In 1852 it sold – in very small quantities – for $545 a pound.[41] The metal began its commercial career in 1886 only after the simultaneous (and totally independent) success of Charles Martin Hall in the United States and Paul Louis Poussaint Heroult in

[38] One report states that energy consumption has been reduced "to about 7.5 kilowatt hours for each pound of aluminum from 12 kilowatt hours a decade ago." *Wall Street Journal,* December 26, 1980, p. 9. The figure of 12 kilowatt-hours for 1970, however, is high by comparison with other estimates. See, for example, The Conference Board, *Energy Consumption in Manufacturing* (Ballinger, Cambridge, Mass., 1974), p. 536.

[39] "One promising new method employs the centuries-old blast furnace techniques used by the iron and steel industry to smelt iron out of iron ore. In this process coal replaces electricity as the principal source of energy. Aluminum ore, silicon and a form of coal are heated in a closed furnace, where the resulting reaction produces molten aluminum silicate, suitable for casting or for further refinement, and carbon monoxide. Alcoa tried this process years ago but abandoned it. Now the increasing value of carbon monoxide as an industrial fuel is beginning to make the process look more worthwhile.

"A big problem with the blast furnace process, however, is that it produces certain aluminum compounds that tend to solidify within the blast furnace, blocking outlets and halting what should be a continuous operation. Alcoa's Mr. Russell says it will be at least a decade before the blast furnace process can be developed on a commercial scale." *Wall Street Journal,* December 26, 1980, p. 9.

[40] In addition to aluminum, other high-performance metals, such as magnesium and titanium, appear also to be locked into electricity-intensive manufacturing technologies.

[41] Peach and Constantin, *Zimmermann's World Resources,* p. 470. "It was so expensive that its use was restricted to such things as the crown jewels and dinnerware for the privileged nobility. Poorer nobility had to struggle along on tableware made of platinum, silver, and gold. Napoleon the Third, the Emperor of France, was so excited about the possibilities of using aluminum for increasing the mobility of his army that he sponsored experiments. Soon thereafter small-scale commercial production began and the price fell rapidly, to $34 a pound in 1856 and to $17 in 1859. Ibid. By the turn of the century, aluminum was selling for thirty-three cents a pound.

France in developing an electrolytic process. The new industry depended totally upon cheap electric power because huge quantities of electricity were required to separate the aluminum from the oxygen in the ore. After the bauxite has been converted into aluminum oxide (alumina), the aluminum oxide "is separated into metallic aluminum and oxygen by direct electric current which also provides the heat to keep molten the cryolite bath in which the alumina is dissolved"[42] The new manufacturing methods became commercially feasible with the availability of cheap electric power at Niagara Falls in the 1890s.[43]

Thus, in its production, aluminum is not only energy-intensive but, more specifically, electricity-intensive. It is highly attractive as an industrial material, however, because it combines high electrical conductivity, high thermal conductivity, and strong resistance to corrosion. Most important for our present purposes, however, is its high strength-to-weight ratio, a characteristic that is even more significant because of the fact that aluminum permits alloying easily and becomes much stronger and stiffer as a consequence. These characteristics offer great opportunities for saving energy. Thus, aluminum has played a major role in transportation equipment, especially automobiles and aircraft, where its lightness provides significant savings in energy.[44] This is especially apparent in automobiles, where the rise in gasoline prices over the past decade and recent federally mandated mileage standards have led to an accelerated use of aluminum. Whereas aluminum use per car was 72.4 pounds in 1964, it had risen to 120 pounds per car in 1979 and is expected to be about 250–285 pounds by 1985.[45] Thus, although aluminum is highly energy-intensive in its manufacture, a complete life-cycle accounting, including the consequences of its industrial applications, would show it to be energy-saving.

Another feature strongly reinforces this point in an energy-accounting context. Aluminum is readily recyclable – again, by the use of electricity-

[42] Charles Carr, *Alcoa* (Rinehart, New York, 1952), p. 86.

[43] Hall's commercial success would not have been possible at a much earlier date when the complementary technology of electric power did not yet exist. Indeed, electrolytic methods for aluminum production had been explored much earlier. "Back in 1854, both Sainte-Claire Deville and Robert Wilhelm Bunsen had explored electrolytic methods of producing aluminum. However, at that time, batteries were the only source of current, and the cost was high. Although Hall still used batteries in his experiments, the dynamo had already been developed." Philip Farin and Gary Reibsamen, *Aluminum* (Metals Week, McGraw-Hill, New York, 1969), p. 11.

[44] The transportation sector absorbs about 20 percent of aluminum production.

[45] *Technological Trends in Major American Industries*, Bulletin no. 1474 (U.S. Department of Labor, Washington, D.C., 1966), p. 85, and U.S. Department of Commerce, Industry and Trade Administration, *1980 U.S. Industrial Outlook*, p. 187.

intensive methods. This recycling is extremely energy-saving. It is estimated that a recycling of secondary aluminum is accomplished in ways that save 95 percent of the energy consumed in producing aluminum from the original bauxite.[46] The electric furnace is the workhorse of the recycling process in various primary metal industries. This has probably been a factor in the growth of electricity utilization elsewhere. It should be apparent that such growth in electricity consumption may indicate energy-saving applications for the economy as a whole. Because of the energy-saving nature of recycling, it is possible that further rises in energy costs will cause increased electricity-intensive recycling in order to bypass the highly energy-intensive mining and processing operations in metallurgy. It is also possible that the ease with which a material can be recycled will emerge as an essential characteristic in a period of high energy costs.

The discussion of electricity so far has been primarily concerned with its specific features in industrial contexts, where it serves as a source of process heat in the chemical transformation of materials. These applications, although vital to industrial economies, have occurred in only a few sectors. For most of the economy, electricity as an energy form has been associated with the introduction of electrically powered machinery. This development began in the last decade of the nineteenth century with the completion of the hydropower complex at Niagara Falls.[47] However, the widespread application of electric power to industry really began after the turn of the century with the perfection of the steam turbine, the central power station, and the electric motor. Although the electric motor accounted for slightly less than 5 percent of mechanical horsepower in 1899, it was over one-quarter of the total just ten years later. By 1919 the figure was 55 percent, by 1929 it was over 82 percent, and by 1939 it reached its peak of just under 90 percent. (see Table 1).

The speed with which electricity became the overwhelmingly dominant power source for industry strongly suggests that this new energy form had compelling advantages. For one thing, electricity could be packaged in any size, whereas steam engines become highly inefficient below a certain size. Electric motors were introduced of just the right

[46] U.S. Department of Commerce, Industry, and Trade Administration, *1980 U.S. Industrial Outlook*, p. 191. Earlier it is pointed out: "Aluminum recovered from old and new scrap represents 23 percent of total aluminum supply. New scrap provides the largest share of the secondary aluminum market. The growth of aluminum in consumer goods is expected to increase old scrap's share of total secondary aluminum supply." P. 170.

[47] See Robert Belfield, "The Niagara System: The Evolution of an Electric Power Complex at Niagara Falls, 1883–1896," Proceedings of the *IEEE*, September 1976.

Table 1. *Electric motor use in relation to total mechanical horsepower in manufacturing, selected years, 1899–1954*

	Total hp (thousand) (1)	Electric motors[a] (thousand hp) (2)	Electric motors as % of total hp (3)
1899	9,811	475	4.8
1904	13,033	1,517	11.6
1909	18,062	4,582	25.4
1914	21,565	8,392	38.9
1919	28,397	15,612	55.0
1925	34,359	25,092	73.0
1929	41,122	33,844	82.3
1939	49,893	44,827	89.8
1954	108,362	91,821	84.7

[a]Represents electric motors driven by purchased electricity and by electric power generated at the establishment.
Source: U.S. Bureau of the Census, *U.S. Census of Manufactures: 1954*, vol. 1 (USGPO, Washington, D.C., 1957), Table 1, p. 207–2.

capacity for each type of industrial application. "Fractionalized" power meant large energy and capital savings since it was no longer necessary to rely upon very large steam engines generating excessive amounts of power, when all that was required were small or intermittent doses. Electricity offered the opportunity of fine-tuning the supply of power to specific needs. Furthermore, the electric motor drastically reduced requirements for floor space and offered increased freedom in the organization and layout of the workplace. The productivity improvements of the assembly-line and mass-production technologies generally owed a great deal to this highly flexible power source.[48]

What is particularly impressive about the shift to electricity as a preferred energy form is the pervasiveness and persistence of this trend. Growing reliance upon electricity is extremely broad-based, suggesting that there are across-the-board factors at work, in addition to those electricity-using innovations, such as the electric arc furnace, that may

[48] As Du Boff has pointed out: "Shortly after steam power began to yield to electricity, installation of electric motors called attention to the obvious restraints placed upon efficiency by the steam engine. Its systems, practices, and factory organization became almost visibly redundant. Thus, as 'unit drive' electric power grew in plant after plant, thoroughgoing reorganization of factory layout and design took place. Machines and tools could now be put anywhere efficiency dictated, not where belts and shafts could most easily reach them." Richard B. Du Boff, "The Introduction of Electric Power in American Manufacturing," *Economic History Review*, December 1967, p. 513.

affect only a small group of industries. Moreover, this trend appears to have continued, at least through most of the 1970s, both in the high-energy-using industries and in the economy as a whole (see Table 2).

It seems obvious that there has been a very wide range of labor-saving innovations throughout industry which have taken an electricity-using form. As a consequence, greater use of electricity is historically the other side of the coin of a labor-saving bias in the innovation process. Such innovations, which increase the productivity of labor in innumerable ways, tend to use electricity because of the great flexibility and convenience apparently associated with that energy form.[49] It is difficult to find a small number of descriptive categories in which to capture these electricity-using innovations. Surveys of innovation in the years since the Second World War continuously refer, in individual industries, to new methods of materials handling, new and very portable tools and equipment, new methods of quality control, automatic lubrication systems, automatic cleaning devices, new techniques for measuring, sensing, detecting, analyzing, and so on.[50] Especially widespread are new methods of automation or continuous-process methods that have substituted for older, small-batch technologies. An additional factor in recent years has been pollution-abatement devices or relatively "clean" electricity-powered technologies that have replaced comparatively "dirty" nonelectrical technologies. Air conditioning has been particularly important in certain industries where delicate equipment and instrumentation or experimental conditions require precise temperature control. However, it has also been widely adopted in many industries where it was believed to increase labor productivity by improving the comfort of the workplace.[51]

We are a long way from being able to deal with these phenomena in a serious quantitative way. For present purposes, I wish to emphasize only that there are numerous compelling reasons for preferring one energy form over another. Energy models that do not take these reasons into account will be very inadequate guides to future energy policy making.

[49] For the first few decades of the twentieth century, see the valuable study by Harry Jerome, *Mechanization in Industry* (National Bureau of Economic Research, New York, 1934).

[50] See, for example, U.S. Department of Labor, *Technological Trends in Major American Industries*.

[51] Anne P. Carter, *Structural Change in the American Economy* (Harvard University Press, Cambridge, Mass., 1970), pp. 54–5.

Table 2. *Purchased energy used by high-energy-using groups distributed by source, selected years, 1967–76 (percentage distribution)*

Industry group (SIC number and title)	Year	Fuels and electricity (1)	Purchased electricity (2)	All fuels (3)	Fuel oil (4)	Natural gas (5)	Coal and coke (6)	Other fuel (7)
All manufacturing	1976	100	38	62	11	35	9	6
	1975	100	37	63	10	36	9	8
	1974	100	36	64	10	37	9	9
	1971	100	31	69	9	39	11	10
	1967	100	28	72	7	35	15	15
26 Paper and allied products								
	1976	100	28	72	31	23	13	4[a]
	1975	100	27	73	30	25	13	5
	1974	100	26	74	28	27	13	5
	1971	100	23	77	25	31	15	5
	1967	100	19	81	18	26	24	12
28 Chemicals and allied products								
	1976	100	36	64	8	42	8	6
	1975	100	35	65	7	42	9	7
	1974	100	33	67	7	44	9	7
	1971	100	30	70	6	42	14	9
	1967	100	30	70	4	38	16	12
29 Petroleum and coal products								
	1976	100	19	81	6	72	Z[b]	3
	1975	100	17	83	6	72	Z	5
	1974	100	16	84	6	68	1	11
	1971	100	13	87	5	74	Z	8
	1967	100	12	88	4	74	1	8

32 Stone, clay, and glass products	1976	100	21	79	10	42	20	7
	1975	100	21	79	9	42	19	9
	1974	100	19	81	8	45	16	11
	1971	100	17	83	8	48	17	10
	1967	100	14	86	5	46	22	13
33 Primary metal industries	1976	100	45	55	9	28	15	3
	1975	100	44	56	9	29	13	5
	1974	100	44	56	8	29	13	5
	1971	100	38	62	7	34	17	5
	1967	100	34	66	7	35	18	6
All other groups	1976	100	48	52	11	26	5	10
	1975	100	47	53	10	26	5	11
	1974	100	46	54	9	28	5	12
	1971	100	41	69	9	27	7	16
	1967	100	36	64	7	19	11	27

[a]Includes coke.
[b]Z represents a figure less than 0.5.
Source: Census of Manufactures: Annual Survey of Manufactures. Reprinted with permission from John G. Myers and Leonard Nakamura, *Saving Energy in Manufacturing* (Ballinger, Cambridge, Mass., 1978), pp. 32–3.

5 On technological expectations

I

One of the most important unresolved issues in the theory of the firm and in the understanding of productivity growth is the rate at which new and improved technologies are adopted.[1] I will argue that expectations concerning the future course of technological innovation are a significant and neglected component of these issues, inasmuch as they are an important determinant of entrepreneurial decisions with respect to the adoption of innovations.

Recent work in various aspects of economic theory and measurement has pointed, once again, to the important role played by expectations of future change in influencing the behavior of economic agents. In a number of instances the expectation of future changes has led to quite different patterns of behavior than might have been expected to have taken place if it had been anticipated that no changes in level and/or trend would occur. What would appear to be aberrational or irrational behavior on the basis of such expectations of no changes is often fully explained once allowance is made for a different set of expectations about the future.

This paper was originally published in the *Economic Journal*, 86, September 1976, pp. 523–35. Reprinted by permission of the Royal Economic Society.

I am heavily indebted to Stanley Engerman for his encouragement and frequent counsel in developing the central argument of this paper and in exploring some of its implications. The paper has also benefited greatly from Paul David's searching criticism of an earlier draft, and from valuable comments and suggestions by David Mowery, W. B. Reddaway, Ed Steinmueller, George Stigler and an anonymous referee. I am also grateful to the National Science Foundation for financial support during the time this paper was being written.

[1] Ed Mansfield's work has forcefully called attention both to the general overall slowness as well as to the wide differentials in adoption rates among different innovations. See, for example, Ed Mansfield, *Industrial Research and Technological Innovation* (W. W. Norton and Co., New York, 1968), chapter 7.

In this paper I should like to draw a similar analogy for the study of the diffusion of technological innovations. The timing and nature of the adoption decision on the part of individual business firms is a key question with major implications for both the micro and macro levels of analysis. I will suggest that there are expectational elements in the adoption decision that have not been given the attention and explored as systematically as could be done to illuminate the diffusion process. Often the explanation of specific rates and patterns of technique adoption seems difficult to comprehend under the implicit conditions assumed, that is, of no future changes in the technological and economic spheres. Yet, as in other parts of economic analysis, the introduction of attention paid to various types of expectations, not only of prices but, more interestingly, of expectations concerning the future rate of technological change itself, will provide some important insights. Since the technological future is, inevitably, shrouded in uncertainty, it is not surprising both that different entrepreneurs will hold different expectations, and also that entrepreneurial behavior will further differ due to varying degrees of risk aversion on the part of decision-makers.

In terms of historical issues there are two patterns of expectations with quite different implications that need to be examined. One, which has been implicit in the comparisons of nineteenth century technological change in the United States and the United Kingdom, is the impact of steady but differing rates of technological change in the two economies. As pointed out by Habakkuk, and more formally by Williamson, the effect of the expectation of higher rates of technological change in the United States should have led to a shorter optimal life for American machinery – that is, in some sense a more rapid introduction of new techniques.[2] More interesting, perhaps, are other patterns of expectations that may be more important in studying the diffusion process. Specifically, at certain times it may be more plausible to anticipate an acceleration of the rate of technological change. Similarly, there may be situations where large-scale improvements are confidently expected *after* the introduction of some major innovation.[3] In such cases these expectations may lead to a surprising result of making rational a delay

[2] H. J. Habakkuk, *American and British Technology in the Nineteenth Century* (Cambridge University Press, 1962), and Jeffrey Williamson, "Optimal Replacement of Capital Goods: The Early New England and British Textile Firm", *Journal of Political Economy*, Nov./Dec. 1971.

[3] For earlier empirical studies dealing with the life-cycle of specific technological innovations, see Simon Kuznets, *Secular Movements in Production and Prices* (Houghton Mifflin, 1930), chapter 1, and Arthur F. Burns, *Production Trends in the United States Since 1870* (National Bureau of Economic Research, 1934), especially chapter 4.

in the widespread diffusion of the innovation. Therefore, in analysing any historical decision with respect to diffusion, one must be sensitive to the specific nature of the expectations held by entrepreneurs with respect to the future course of technology.

II

If, as Alfred North Whitehead once asserted, the history of western philosophy may be adequately described as a series of footnotes upon Plato, it may equally be said of the study of technological innovation that it still consists of a series of footnotes upon Schumpeter. Although the footnotes may be getting longer, more critical and, happily, richer in the recognition of empirical complexities, we still occupy the conceptual edifice that Schumpeter built for the subject. Inevitably, therefore, Schumpeter's concepts constitute our point of departure.

Schumpeter's theory of capitalist development, it will be recalled, starts out from the circular flow of economic life where producers and consumers are all in equilibrium, and where all adjustments and adaptations have been made. Schumpeter then introduces an innovation – a shift in the production function – into this circular flow. The entrepreneurial response to this new profit prospect in turn generates a sequence of alterations in the behavior of economic actors, beginning with an expansion of bank credit and including, eventually, a secondary wave of investment activity imposed on top of the primary wave as the expectations of the larger business community are affected by the evidence and by the consequences of business expansion.[4]

Schumpeter himself was so much persuaded of the large elements of risk and uncertainty inherent in the innovation decision that he downplayed the role of rational calculation itself in the decision-making process. The Schumpeterian entrepreneur is a distinctly heroic figure, prepared (unlike most mortals) to venture forth boldly into the unknown. His decisions are not the outcome of precise and careful calculation and, Schumpeter emphasised, cannot be reduced to such terms.

The point to be made here is that there is a further dimension of uncertainty in the innovation decision of a sort not emphasised by Schumpeter in his stress on the *discontinuous* nature of technological innovation. This is, quite simply, the uncertainty generated not only by technological innovations elsewhere in the economy but *by further improvement in the technology whose introduction is now being considered.* Schumpeter's argument creates a presumption that the first

[4] J. A. Schumpeter, *The Theory of Economic Development* (Harvard University Press, 1934), chapters 1 and 2.

innovator reaps the large rewards. Nevertheless, the decision to undertake innovation *X* today may be decisively affected by the expectation that significant improvements will be introduced into *X* tomorrow (or by the firmly held expectation that a new substitute technology, *Y*, will be introduced the day after). The possible wisdom of waiting is reinforced by observations, abundantly available to all would-be entrepreneurs, concerning the sad financial fate of innumerable earlier entrepreneurs who ended up in the bankruptcy courts because of their premature entrepreneurial activities.[5] As soon as we accept the perspective of the ongoing nature of much technological change, the optimal timing of an innovation becomes heavily influenced by expectations concerning the timing and the significance of *future* improvements. Even when a new process innovation passes the stringent test of reducing new average total costs below old average variable costs, it may not be adopted. The reason for this is that the entrepreneur's views about the pace of technological improvements may reflect expectations of a higher rate of technological obsolescence than that allowed for by conventional accounting procedures in valuing the investment. Moreover, accounting formulae may not give adequate recognition to the "disruption costs" involved in introducing new methods, especially when such disruptions are frequent. Thus, a firm may be unwilling to introduce the new technology if it seems highly probable that further technological improvements will shortly be forthcoming.[6] This problem of how the optimal timing of innovation is

[5] Marx long ago called attention to "the far greater cost of operating an establishment based on a new invention as compared to later establishments arising *ex suis ossibus*. This is so very true that the trail-blazers generally go bankrupt, and only those who later buy the buildings, machinery, etc., at a cheaper price, make money out of it" (Karl Marx, *Capital* [Foreign Languages Publishing House, Moscow, 1959], vol. III, p. 103). He also called attention to the rapid improvements in the productivity of machinery in its early stages as well as the sharp reduction in the cost of its production. "When machinery is first introduced into an industry, new methods of reproducing it more cheaply follow blow by blow, and so do improvements, that not only affect individual parts and details on the machine, but its entire build" (Karl Marx, *Capital* [Modern Library Edition, New York, no date], vol. I, p. 442). In a footnote on that page, Marx cites approvingly Babbage's statement: "It has been estimated, roughly, that the first individual of a newly invented machine will cost about five times as much as the construction of the second." For a discussion of related problems with respect to the growth of nations, see Ed Ames and Nathan Rosenberg, "Changing Technological Leadership and Economic Growth", *Economic Journal*, March 1963.

[6] Fellner's discussion of what he calls "anticipatory retardation" is relevant here. See William Fellner, "The Influence of Market Structure on Technological Progress", *Quarterly Journal of Economics*, 1951, pp. 556–77, as reprinted in R. Heflebower and G. Stocking (eds.), *Readings in Industrial Organization and Public Policy* (Richard D. Irwin, 1958). The discussion of "anticipatory retardation" appears on pages 287–8 of that volume.

affected by expectations of discontinuous technological change is an extremely significant one that has received relatively little attention in the theoretical literature.[7] Nor have there been systematic empirical studies of the phenomenon.

In their earliest stages, innovations are often highly imperfect and are known to be so. Innumerable "bugs" may need to be worked out.[8] If one anticipates significant improvements, it may be foolish to undertake the innovation now – the more so the greater the size of the financial commitment and the greater the durability of the equipment involved. Whereas the Schumpeterian innovator experiences abnormally high profits until the "imitators" catch up with him, the impetuous innovator may go broke as a result of investing in a premature model of an invention. This apparent distinction follows from the earlier-stated diffi-

[7] Of course, attention has been paid to the optimal timing of the introduction (and scrapping) of machinery, but these models generally do not deal with problems of expected future changes in technology. Rather, they are more often concerned with issues relating to relative factor prices, future product demands, or the relation between machine use and deterioration. While the expectations relating to technology can no doubt be easily incorporated into such models, the specific problem is virtually unexplored. For a brief discussion, see Vernon L. Smith, *Investment and Production* (Harvard University Press, 1961), pp. 143–5. The relationship between expectations concerning the rate of technological progress and the rate of return on investment in the context of a model of embodied technological change is discussed by Robert Solow in *Capital Theory and the Rate of Return* (North Holland Publishing Co., 1963, pp. 61–4). For an interesting treatment of a somewhat related problem, how the introduction of a new technology will be affected by expected future growth in demand under different forms of market structure, externalities and property rights, see Yoram Barzel, "Optimal Timing of Innovations", *Review of Economics and Statistics,* August 1968.

[8] This term should be taken to include a great many production problems involving the use of new equipment that become apparent only as a result of extensive use – for example, metal fatigue in aeroplanes. William Hughes has made this point well with respect to exploration of the scale frontier in electric power generation: "Even under the most favorable conditions for advancing the scale frontier the cost side of the equation imposes fairly strict upper limits on the economical pace of advance, and trying to force the pace could mean sharply rising cost of development. The experience required for pushing out the scale frontier is related to time and cannot be acquired by increasing the number of similar new units. Perhaps the greatest uncertainties connected with units arise from problems that may not show up until the units have been in operation a few years. For the industry as a whole, the socially optimal number of pioneering units during the first two or three years of any major advance in scale, design, or steam conditions is probably rather small, most often ranging from perhaps two or three or half a dozen" (William Hughes, "Scale Frontiers in Electric Power," in William Capron [ed.], *Technological Change in Regulated Industries* [The Brookings Institution, Washington, D.C., 1971], p. 52). One of the other virtues of the Hughes article is its forceful reminder of the intimate link that often exists between technological progress and economies of scale. "The realization of latent scale economies is an especially important form of technological progress in the utility industries" (ibid., p. 45).

culties of Schumpeter's concept of innovation with its emphasis on discontinuity and its implication that all problems in the introduction of a new product or process have already been completely solved. Moreover, as Mansfield points out: "In cases where the invention is a new piece of equipment, both the firm that is first to sell the equipment and the firm that is first to use it may be regarded as innovators. The first user is important because he, as well as the supplier, often takes considerable risk."[9]

De Tocqueville long ago pointed out a distinctive characteristic of the American scene: "I accost an American sailor, and I inquire why the ships of his country are built so as to last but for a short time; he answers without hesitation that the art of navigation is every day making such a rapid progress that the finest vessel would become almost useless if it lasted beyond a certain number of years. In these words, which fall accidentally and on a particular subject from a man of rude attainments, I recognize the general and systematic idea upon which a great people directs all its concerns."[10] Similarly, expectations of future changes may have the effect, not of delaying innovation, but of determining some of the specific characteristics of the innovation chosen. An adaptation to these expectations might be deliberately to construct cheaper and flimsier capital equipment. Thus, for example, as between two economies with different expected rates of future technical change, it would be anticipated that optimal life would be shorter where expected changes are largest.[11] And, correspondingly, an anticipated acceleration in the course of technological improvement could lead to the selection of equipment with expected optimal life shorter than would otherwise be chosen.

Central to the analysis of the problem of expectations with respect to further improvements in technology is the question of the specific source of subsequent improvements. It may be that these improvements can only be brought about by the process which we have come to call "learning by doing", in which case the pace of improvement is itself determined by either the producers' accumulated experience over time or by their cumu-

[9] Edwin Mansfield, *Industrial Research and Technological Innovation*, p. 83.

[10] Alexis De Tocqueville (trans. Henry Reeve), *Democracy in America*, 2 vols. (D. Appleton and Company, New York, 1901), vol. I, p. 516.

[11] For a discussion of optimal replacement in the *ante bellum* textile industries of the United States and the United Kingdom, see Williamson, op. cit. For a criticism of the Williamson article that does not alter the relationship between expected rates of technical change and optimal life, see David Denslow, Jr., and David Schulze, "Optimal Replacement of Capital Goods in Early New England and British Textile Firms: A Comment", *Journal of Political Economy*, May/June 1974.

lative output.[12] Alternatively, improvements may be rigidly linked to the passage of time necessary to acquire information about the results of earlier experience.[13] Consider the problems associated with the introduction of commercial jet aeroplanes. Britain introduced the Comet I two years before the Americans began the development of a jet airliner. Yet the Americans eventually won out. In retrospect, it is apparent that the American delay was salutary rather than costly to them, and that Boeing and Douglas chose the moment to proceed better than did de Havilland. "Their delay allowed them to offer airplanes that could carry up to 180 passengers when the Comet IV carries up to 100, and a cruising speed of 550 mph instead of 480 mph – hard commercial advantages that they could offer because they were designing for later and more powerful engines. But they were also aided by the delay of four years in making the Comet safe after its accidents from metal fatigue."[14] More generally, information concerning the useful life of metal components could only be derived from prolonged periods of use and experience. Which forms of improvements may be expected to dominate, and the conditions under which they may become available to other firms, will have important

[12] See Paul David, "Learning by Doing and Tariff Protection: A Reconsideration of the Case of the Ante-Bellum United States Cotton Textile Industry", *Journal of Economic History,* September 1970, and Tsuneo Ishikawa, "Conceptualization of Learning by Doing: A Note on Paul David's 'Learning by Doing and . . . The Ante-Bellum United States Cotton Textile Industry' ", ibid., December 1973. Note the importance of this distinction as well as that between learning that can be freely captured by others and that which accrues solely to the learner. This latter distinction will have important implications for the number of firms in an industry as well as the optimal entry date for any one firm. See Barzel, op. cit.

[13] This is not really "learning by doing" in the usual sense, which is restricted to learning by participation in the production process, even though it does involve the passage of time and the accumulation of experience. However, the distinctions above concerning the "appropriability" of the information generated by the experience remain important for examining entry decisions.

[14] Ronald Miller and David Sawers, *The Technical Development of Modern Aviation* (Praeger Publishers, New York, 1970), p. 27. For an American failure in an attempted development of the "Demon" fighter plane under the pressures of the Korean War, see Eighth Report of the Preparedness Investigating Sub-committee of the Committee on Armed Services, United States Senate (84th Congress, 2d Session), *Navy Aircraft Procurement Program: Final Report on F3H Development and Procurement* (United States Government Printing Office, Washington, 1956). The specific failure here was that numerous airframes became available years before it was possible to develop a jet engine with the required performance characteristics. In the lugubrious tone of the congressional investigating committee that was appointed to account for the resulting loss of several hundred million dollars: "What has somehow been overlooked is that the essential procurement practice employed with respect to the Demon fighter would inevitably result in some wastage of airframes if the engine were not forthcoming" (ibid., pp. 9–10).

implications not only for entrepreneurs making entry or adoption decisions, but also for public policy concerning efficient growth.

There are many possible reasons why waiting may be the most sensible decision. Indeed, often there may be no real choice. On the purely technological level, innovations in their early stages are usually exceedingly ill-adapted to the wide range of more specialised uses to which they are eventually put. Potential buyers may postpone purchase to await the elimination of "bugs" or the inevitable flow of improvements in product performance or characteristics. On the other hand, they may have to wait through the lengthy process of product redesign and modification before a product has been created that is suitable to specific sets of final users. Thus, widely used products such as machine tools, electric motors and steam engines have experienced a proliferation of time-consuming changes as they were adapted to the varying range of needs of ultimate users. In the case of machine tools, for example, there was a successful search for the application of specific machine tool innovations across a wide variety of industrial uses.[15] In the case of final consumer goods, the redesigning is likely to be primarily concerned with the development of product varieties suitable to the financial resources of different income groups. What is observed over their life cycles is a gradual expansion of their quality range to accommodate these final users.[16] Today's academics are keenly aware of this problem when deciding whether to purchase today or to defer the purchase of pocket calculators, given their expectations of ongoing improvement in their capacities and characteristics. Similar problems have characterised the selection decision with respect to the last few generations of computers, and may also have influenced the choice between purchase and rental on the part of users.[17]

There are some areas, of course, where the pressures militate very heavily against delay – as in weapons acquisition. Nevertheless, even when the costs of being late may be regarded as uniquely high, considerations of technological expectations cannot be ignored. Peck and Scherer, in their very careful study of weapons procurement, rephrase two of the key questions. First, "Should we begin developing this newly feasible weapon system immediately to ensure early availability, or should we wait a year or so until the state of the art is better defined so

[15] Nathan Rosenberg, "Technological Change in the Machine Tool Industry, 1840–1910", *Journal of Economic History*, December 1963.

[16] For historical evidence in support of these assertions, see Dorothy Brady, "Relative Prices in the Nineteenth Century", *Journal of Economic History*, June 1964, pp. 145–203.

[17] See William F. Sharpe, *The Economics of Computers* (Columbia University Press, New York, 1969), particularly chapter 7.

that fewer costly mistakes will be made during the development?" Secondly, "should funds be committed to long lead time production items early in the development program so that full-scale production can start as soon as the development is completed?"[18]

A recent important instance in which the combination of long lead time and rapid technological change led to an apparent large misallocation of resources is the case of automobile emission controls. Under pressure from the federal government, American automakers, with production lead times of four or five years, were required to specify which anti-pollution technology would be incorporated into the 1975 automobiles. The catalytic converter seemed at that time to be a proven and more certain solution to the problem than did alternative technologies, such as the still-unproven stratified-charge engine. However, the subsequent development of this latter technology produced a solution to automobile emissions that was superior to the catalytic converter. Foreign automakers, such as Honda, with shorter production lead times than the Americans, were able to adopt the stratified-charge technology, thus avoiding what now appears to be the costly and premature commitment of American auto firms. (It is important to note, however, that some of the difficulty in this particular case is due to uncertainty concerning the future of the emission standards themselves as well as technological uncertainty.)

Problems such as these have, of course, a long history. In his book on the early history of electrical equipment manufacturing, Harold Passer extensively documents the marketing difficulties that confronted makers before 1900. During this period, expectations of rapid technical improvement were firmly entrenched in the minds of potential buyers, and such expectations served to reduce present demand. "The manufacturer has to convince the prospective buyer that no major improvements are in the offing. At the same time, the manufacturer must continue to improve his product to maintain his competitive position and to force existing products into obsolescence."[19] Passer's statement neatly identi-

[18] Merton J. Peck and Frederic M. Scherer, *The Weapons Acquisition Process: An Economic Analysis* (Division of Research, Graduate School of Business Administration, Harvard University, Cambridge, 1962), pp. 283, 318. The authors also observe: "The risks of early investment in production are impressive – in some programs enormous quantities of special tools and manufactured material became worthless due to unexpected technical changes. On the other hand, the gamble has been successful and valuable time savings have been achieved in other programs (such as in the prewar British radar effort and in the first U.S. ballistic missile programs) by preparing for production concurrently with development" (ibid., pp. 318–19).

[19] Harold Passer, *The Electrical Manufacturers, 1875–1900* (Cambridge, Harvard University Press, 1953), p. 45. The 1880s and the 1890s were, in fact, a period of continuous

fies the horns of the dilemma that threaten to impale the entrepreneur in the early stages of product innovation. One must attempt to persuade potential buyers of product stability at the same time as one commits resources to the search for product improvement so as not to fall behind the pace of such improvement that is set by one's competitors.

The implications of such considerations may be very great for the innovation process. When these intertemporal considerations loom very large, a rapid rate of technological progress need by no means result in as rapid a rate of introduction of new technological innovation. As Sayers has pointed out for Great Britain:

> There were times, between the wars, when marine engineering was changing in such a rapid yet uncertain way that firms in the highly competitive shipping industry delayed investment in the replacement of old high-cost engines by the new low-cost engines. In the middle 'twenties progress was rapid in all three propulsion methods – the reciprocating steam-engine, the geared steam turbine and the diesel motor. Minor variations are said to have brought the number of possible combination types up to nearly a hundred. For some classes of ships there was momentarily very little to choose between several of these combinations, and shipowners were inclined to postpone placing orders until a little more experience and perhaps further invention had shown which types would be holding the field over the next ten years. Put in economic terms, the shipowners' position was that, though total costs of new engines might already be less than running costs of old engines, the profit on engines of 1923 build might be wiped out by the appearance in 1924 of even lower-cost engines, the purchase of which would allow a competitor (who had postponed the decision) to cut freights further. Also there was uncertainty as to which of two types of 1923 engine would prove to work at lower cost. If shipowner *A* installed engine *X*, and shipowner *B* installed engine *Y*, whose costs in 1923 appeared to equal those of *X*, a year's experience might show that in fact *Y* costs were much lower than *X* costs, in which event

product improvement in incandescent lighting, and it was only with the advent of metallic filaments in the early years of the twentieth century that the incandescent lamp established its decisive superiority over gas lighting and arc lighting. Even so, the record from 1880 to 1896 is one of startling improvements in product efficiency and reductions in cost. Bright has stated: "The following figures show the approximate course of list prices per lamp for standard 16-candlepower lamps from 1880 to 1896: 1880–6, $1.00; 1888, 80 cents; 1891, 50 cents; 1892, 44 cents; 1893, 50 cents; 1894, 25 cents; 1895, 18–25 cents; 1896, 12–18 cents . . . In 1896 a dollar could buy approximately six standard carbon-filament lamps which would give more than twelve times as many lumen-hours of light as a single lamp of the same candlepower costing a dollar in 1880" (Arthur A. Bright, Jr., *The Electric-Lamp Industry* (New York, MacMillan, 1949), pp. 93 and 134).

> shipowner *A* would have done better to wait until 1924 before installing new engines.[20]

For a much more recent period, Walter Adams and Joel Dirlam call attention to the U.S. Steel Corporation's concern over the imminence of future improvements as an explanation for their delay in introducing the oxygen steel-making process.

> U.S. Steel conceded that "some form of oxygen steel-making will undoubtedly become an important feature in steelmaking in this country", but it declined to say when or to commit itself to introducing this innovation. Indeed, three years later, *Fortune* still pictured the Corporation as confronted by "painfully difficult choices between competing alternatives – for example, whether to spend large sums for cost reduction *now* [1960], in effect committing the company to *present* technology, or to stall for time in order to capitalise on a new and perhaps far superior technology that may be available in a few years".[21]

Expectations of continued improvement in a new technology may therefore lead to postponement of an innovation, to a slowing down in the rate of its diffusion, or to an adoption in a modified form to permit greater future flexibility. Moreover, one must consider expectations relating not only to possible improvements in the technology being considered but also the possibility of improvement in both substitute and complementary technologies. Further improvements in an existing product may be held up because of the expectation that a superior *new* product will soon be developed.[22] At the same time expectations of continued improvement in the *old* technology, which the new technology is designed to displace, will exercise a similar effect. There is much evidence to suggest that historically, the *actual* improvement in old technologies after the introduction of the new were often substantial

[20] R. S. Sayers, "The Springs of Technical Progress in Britain, 1919–39", *Economic Journal*, June 1950, pp. 289–90. These circumstances might also be described as a case of "technological uncertainty" as to which specific methods would yield the greatest long-run effect, but the impact upon diffusion is the same. Habakkuk has called attention to a similar experience in shipbuilding immediately after the opening of the Suez Canal: "It accelerated the technical perfection of the steamer; the rate of technical progress was so rapid that the steamers built in the early 'seventies were unable to compete with those completed in the middle of the decade" (H. J. Habakkuk, "Free Trade and Commercial Expansion, 1853–1870", in *The Cambridge History of the British Empire*, vol. 2, p. 762).

[21] Walter Adams and Joel Dirlam, "Big Steel, Invention, and Innovation," *Quarterly Journal of Economics*, May 1966, pp. 181–2. Emphasis by Adams and Dirlam.

[22] Jewkes et al. have pointed out that "improvements in the more traditional methods of producing insulin were held up by the widespread belief that a synthesized product would soon be found." See John Jewkes, David Sawers and Richard Stillerman, *The Sources of Invention* (Macmillan and Co., London, 1958), p. 232.

and played a significant role in slowing the pace of the diffusion process, so that this provides a quite reasonable basis for such a set of expectations. The water wheel continued to experience major improvements for at least a century after the introduction of Watt's steam engine; and the wooden sailing ship was subjected to many major and imaginative improvements long after the introduction of the iron-hull steamship.[23] During the 1920s the competition of the internal combustion engine is said to have been responsible for much technological improvement in steam engines, while in the same period the competition from the radio stimulated experiments that led to the new and improved type of cable that was introduced in 1924.[24] The Welsbach mantle, perhaps the single most important improvement in gas lighting, was introduced after the electric utilities had begun to challenge the gas utilities over the respective merits of their lighting systems. The Welsbach gas mantle brought about a dramatic increase in the amount of illumination produced by a standard gas jet.[25] Not only the diffusion of technologies but also the effort devoted to the development of new technologies may be decisively shaped by expectations as to future improvements and the continued superiority of existing technologies. One explanation for the limited attention devoted to the development of the electric motor for many years was the belief that the economic superiority of the steam engine was overwhelming and beyond serious chal-

[23] See Nathan Rosenberg, "Factors Affecting the Diffusion of Technology", *Explorations in Economic History,* Fall 1972, for a more extended discussion.

[24] Sayers, op. cit., pp. 284–5.

[25] "The Welsbach mantle was a lacy asbestos hood which became incandescent when attached over a burning gas jet. It increased the candlepower six-fold with a white light superior to the yellowish flame of the bare jet . . . The Welsbach mantle, improved by many changes in the original design, extended the life of gas lighting nearly half a century. It sustained the gas utilities while they were discovering other productive uses for gas which would permit them to give up the struggle against electric lighting" (Charles M. Coleman, *P.G. and E. of California,* the Centennial Story of Pacific Gas and Electric Company, 1852–1952), p. 81.

Even the old arc-lamp technology experienced substantial improvements in response to the competition of new forms of lighting. See Bright, op. cit., pp. 213–18.

It is interesting to note that, even though the gas industry could not indefinitely meet the competition of electricity in lighting, the gas industry as a whole continued to find new uses for its product and experienced no long-term decline. "Despite the declining importance of gas in lighting, the manufactured-gas industry as a whole has grown continually. The competition of the carbon lamp and the old open arc during the 1880s had encouraged the gas industry to spread its field to heating, and it was that use which permitted the industry to continue expanding when its lighting market was destroyed. Cooking, space heating, water heating, and later refrigeration resulted in a steady growth in the value of products of the industry from $56,987,290 in 1889 to $512,652,595 in 1929" (Bright, op. cit., p. 213).

lenge.[26] The decision to neglect research on the electrically powered car in the early history of the automobile industry reflected the belief, justified at the time, in the total superiority of the internal combustion engine (this neglect may soon be repaired!). Similarly, the limited shift to nuclear sources of power over the past quarter century has been influenced by continued improvements in thermal efficiency based upon the "old-fashioned" but still apparently superior fossil fuel technologies. It seems equally clear that the recent growing pessimism about future fossil fuel supplies is likely to accelerate the concern with nuclear technologies, and lead to a more rapid rate of technical improvements there as experience accumulates. This will first require, however, a careful sorting out and distinction between short-term and long-term phenomena. Large-scale commitment of private resources to nuclear energy (or shale processing) is unlikely so long as investors in energy-supplying industries anticipate that the price of oil is likely to fall sharply from its post-October 1973 levels.[27]

The recent efforts by oil firms to gain control of competing technologies – coal and uranium – may be seen as attempts to assure long-term market control by minimising the potential threats arising from technological breakthroughs in the provision of substitute products.[28] The earlier experience of gas companies in meeting the competition of electricity is also highly instructive in this regard. The ubiquitousness of "Gas and Electric" utility companies in the United States today suggests something of their past success in making the transition from the old to the new technology.

Expected profitability will not only be affected negatively by expected improvements in substitute technologies; it will also be affected positively by expected technological improvements in complementary technologies. Since innovation in steel-making will be affected not only by innovations in aluminum or pre-stressed concrete which provide potential substitutes for steel but also by technological changes in petroleum

[26] Kendall Birr, "Science in American Industry", in David Van Tassel and Michael Hall (eds.), *Science and Society in the Untied States* (The Dorsey Press, Homewood, Illinois, 1966), p. 50.

[27] Dependence upon price expectations has, of course, been extensively discussed elsewhere. The purpose of the present article is confined to calling attention to the significance of expectations with respect to technological change itself. Expectations about future changes in technology might themselves be produced by projecting the anticipated influence of relative price changes on the innovation process. The willingness to explore for new techniques that involve factor combinations drastically different from those that currently prevail, will in turn depend upon the magnitude of expected price shifts.

[28] Such attempts, by providing a limited diversification for these firms, reduce the risks they face, and this may conceivably have some role in encouraging research activity.

refining which increase the size of the automobile-producing sector, the entrepreneur must consider expected developments in these other sectors. The profitability of technological changes in electric power generation will be favourably affected by metallurgical improvements that provide power plant components with increased capacity to tolerate higher pressures and temperatures in the form of high-strength, heat-resistant alloys. On the other hand, improvements in power generation would have a limited impact upon the delivered cost of electricity without improvements in the transmission network that reduce the cost of transporting electricity over long distances. The point is that the *need* for and expected availability of complementary innovations will often affect the profitability and therefore the diffusion of an innovation.[29] Therefore single technological breakthroughs hardly ever constitute a complete innovation.[30] It is for this reason – the expected as well as realised changes in other sectors – that decisions to adopt an innovation are often postponed in situations that might otherwise appear to constitute irrationality, excessive caution, or overattachment to traditional practices in the eyes of uninformed observers.

III

Some significant (and superficially paradoxical) implications can be drawn from this discussion. Specifically, the relationship between the rate of technological change and the rate of technological innovation and diffusion is by no means a simple one, and may well be the opposite of intuitive expectations. A rapid rate of technological change may lead to a seemingly slow rate of adoption and diffusion, or to the introduction of machinery that fails to incorporate the most "ad-

[29] See Fishlow's discussion of the importance of expectations with respect to transport improvements and their impact upon American land settlement patterns. Albert Fishlow, *American Railroads and the Transformation of the Ante-Bellum Economy* (Harvard University Press, 1965). Fishlow explains why mid-western railroads were profitable immediately after their construction as a result of the process of anticipatory settlement. More broadly, it would be of interest to examine the impact of the expectations generated by transport improvements upon agricultural practices in the older areas of the northeast.

Parker and Klein have argued that the mechanisation of American agriculture that so increased its productivity would not have occurred in the absence of those transport improvements that made it possible to introduce its products into world markets. See William Parker and Judith Klein, "Productivity Growth in Grain Production in the United States, 1840–60 and 1900–10," in Studies in Income and Wealth No. 30, *Output, Employment and Productivity in the United States after 1800* (Columbia University Press, 1966).

[30] Rosenberg, "Factors Affecting the Diffusion of Technology", op. cit.

vanced" technology, so long as it leads potential buyers to anticipate, by extrapolation, a continued or accelerating rate of future improvements. A decision to buy now may be, in effect, a decision to saddle oneself with a soon-to-be-obsolete technology.[31] Conversely, when the rate of technological change slows down and the product stabilises, the pace of adoptions may increase owing to the much greater confidence on the part of potential buyers that the product will *not* be superseded by a better one in a relatively short period of time. Thus the lag behind the "best available" methods might appear less when technological change is at a slower rate or is decelerating than when it is more rapid. In this sense, our argument may explain the failure of firms to function at the "best practice" frontier – such failure may owe a great deal to differences in entrepreneurial expectations about the future pace of technological improvement.[32]

There are two distinct issues involved here. One is the rate of improvement of best practice technology. The second is the rate of adoption of those best practice methods. If the two were independent, any lag in adoption would seem to impose social costs. However, as seems clear from historical experience, an important explanation of lagged adoption is the environment created by the high rate of improvement of best practice technology. Thus, a lagged rate of adoption is the "price" paid by technologically dynamic economies for their technological dynamism.

Clearly, a further examination of the adoption of technological innovations must be conducted within an enlarged framework that includes expectations not only of own-improvements but of improvements in the range of closely linked substitutes and complements as well. Further research along such lines may considerably improve our understanding of the diffusion process. For the present, I would suggest that decisions to postpone the adoption of an innovation are often based upon well-founded and insufficiently appreciated expectations concerning the future time-flow of further improvements. Even the most widely accepted justification for postponement, the elimination of conspicuous but not

[31] As discussed, the effect might also be to lead to the initial adoption of a more flexible set of techniques that more readily permits the adoption of future improvements. For an analogous argument with respect to the benefits of flexibility in the design of plant and equipment, see George Stigler, "Production and Distribution in the Short Run", *Journal of Political Economy,* June 1939. Reprinted in American Economic Association, *Readings in the Theory of Income Distribution* (Philadelphia, 1946).

[32] For a discussion of other reasons for the failure to operate at the "best practice" frontier, see W. E. G. Salter, *Productivity and Technical Change* (Cambridge University Press, 1960), chapter 4.

overwhelmingly serious technical difficulties, or "bugs", can reasonably be interpreted as merely a special case of expectations of future technological improvement. In this case, the expectations approach complete certainty that the technical difficulties can shortly be eliminated by recourse to the application of ordinary engineering skills. I suggest, finally, that entrepreneurs may be making appraisals of the future payoff to innovations of greater objective validity than are made by social scientists who invoke all sorts of extra-rational factors to account for the delay or "lag" in the adoption and diffusion of innovation. Practical businessmen tend to remember what social scientists often forget: that the very rapidity of the overall pace of technological improvement may make a postponed adoption decision privately (and perhaps even socially) optimal.

6 Learning by using

The assertion that economic activity involves a significant learning dimension would probably command universal agreement. Indeed, at that level of bald and unqualified assertion, there is very little to disagree with. The more interesting questions arise when we attempt to give the assertion some empirical or analytical content.

I will argue that we may fruitfully look upon technological innovation as a learning process – in fact, as several distinct kinds of learning processes. Sometimes, as we will see, these overlap or feed back upon one another. An essential first step, then, is to recognize that several different categories of learning exist. It is important to achieve some conceptual clarity because, as I will argue, there is at least one form of learning in the innovation process that has not been recognized and that is probably of increasing significance in a high-technology economy. This paper should therefore be regarded as part of a larger intellectual enterprise whose purpose is to identify the several different types of learning processes, as well as the nature of these processes and the ways in which they feed into the larger pattern of activities that constitute technological innovation. The main focus is not upon major innovations of Schumpeterian magnitude but, rather, upon the minor improvements that determine the rate of productivity growth that major innovations are capable of generating. In this respect, one main purpose of the paper is to clarify the linkages between technological processes and their economic consequences.

What we now call "research and development (R&D)" is a learning process in the generation of new technologies. Rather, R&D includes

Sections III and IV draw heavily, in parts, on Nathan Rosenberg, Alexander Thompson, and Steven E. Belsley, "Technological Change and Productivity Growth in the Air Transport Industry," NASA Technical Memorandum 78505, September 1978.

The author has had the considerable benefit of discussions with Stanley Engerman, Marie-Therese Flaherty, David Mowery, Keith Pavitt, and Edward Steinmueller. The research on which this paper is based has been supported by NSF Grant PRA-77-21852.

several different forms of learning that are relevant to the innovation process. The detailed analysis of this potpourri concept is not our primary purpose here, and so it will be dispatched in a few words. We have thrown a variety of very different kinds of activities into this particular pot. At the basic research end of the spectrum, the learning process involves the acquisition of knowledge concerning the laws of nature. Some of this knowledge turns out to have useful applications to productive activity. At the development end of R&D is a learning process that consists of searching out and discovering the optimal design characteristics of a product. At this stage, the learning is oriented toward the commercial dimensions of the innovation process: discovering the nature and combination of product characteristics desired in the market (and in relevant submarkets), and incorporating these in a final product in ways that take into account scientific and engineering knowledge. This is a very subtle process, particularly when, as is increasingly common, we are dealing with products having a high degree of systemic complexity.

The aspects of learning commanding most attention in recent years deal with learning by doing, which Arrow emphasized in his seminal article "The Economic Implications of Learning by Doing."[1] This is a form of learning that takes place at the manufacturing stage after the product has been designed, that is, after the learning in the R&D stages referred to above has been completed. Learning at this stage, as described by Arrow and others, consists of developing increasing skill in production. This has the effect of reducing real labor costs per unit of output. The significance of this process has been documented in several industries, including airframe production, shipbuilding, machine tools, and textiles.[2]

A related form of learning by doing, which is not highlighted in the literature, also occurs as a by-product of productive activity. The point here is that there are many kinds of productivity improvements, often individually small but cumulatively very large, that can be identified as a result of direct involvement in the productive process. This is a source of technological innovation that is not usually explicitly recognized as a component of the R&D process, and receives no direct expenditures –

[1] Kenneth Arrow, "The Economic Implications of Learning by Doing," *Review of Economic Studies*, June 1962.

[2] A. Alchian, "Reliability of Progress Curves in Airframe Production," *Econometrica*, October 1963; Werner Hirsch, "Firm Progress Ratios," *Econometrica*, April 1956; Leonard Rapping, "Learning and World War II Production Functions," *Review of Economics and Statistics*, February 1965; Paul David, "Learning by Doing and Tariff Protection" *Journal of Economic History*, September 1970.

which may be the reason it is ignored. It overlaps with development. The learning involved requires participation in the production process. Such participation is obviously not sufficient, because a perception of possible improvements depends not only upon the opportunity to make certain observations but also upon prior training and experience. The point is that productive activities always involve specialized kinds of knowledge, much of which may be unique to a specific industrial process. There is typically a range of possible improvements that require intimate familiarity with the minutiae of the productive sequence.[3]

II

The categories of learning mentioned so far deal with producing new scientific knowledge, incorporating new knowledge in the design of a new product, learning new productive activities when a novel product is put into production, and learning ways to improve the productive process itself that grow out of experience with that process. I now wish to call attention to a separate category of learning that begins only after certain new products are used. With respect to a given product, I want to distinguish between gains that are internal to the production process (doing) and gains that are generated as a result of subsequent use of that product (using). For in an economy with complex new technologies, there are essential aspects of learning that are a function not of the experience involved in producing the product but of its *utilization* by the final user. This is particularly important in the case of capital goods. Thus, learning by using refers to a very different locus of learning than does learning by doing. There are various reasons why this should be so. Perhaps in most general terms, the performance characteristics of a durable capital good often cannot be understood until after prolonged experience with it. For a range of products involving complex, interdependent components or materials that will be subject to varied or prolonged stress in extreme environments, the outcome of the interaction of these parts cannot be precisely predicted. In this sense, we are dealing with performance characteristics that scientific knowledge or techniques cannot predict very accurately. The performance of these products, therefore, is highly uncertain. Moreover, many significant characteristics of such products are revealed only after intensive or, more significantly, prolonged use. Indeed, one of the central characteris-

[3] For further discussion of the cumulative importance of individually small improvements, see Nathan Rosenberg, "Technological Interdependence in the American Economy," *Technology and Culture*, January 1979, especially pp. 32–40.

tics is the useful life of an expensive asset. One of the basic purposes of this learning-by-using process is to determine the optimal performance characteristics of a durable capital good as they affect the length of useful life. Closely related to this is the fact that optimal servicing and maintenance characteristics can be determined only after extensive use – in many cases, only after many years.

Much of the technical knowledge required in high-technology societies tends to be extremely specialized or specific with respect to the nature of the process and the machines involved. Not only is this knowledge not precisely predictable from scientific principles or methodology; it cannot, as we will see, be accurately predicted from experience with related or analogous technologies.

Another process generating such knowledge has been analyzed by von Hippel and raises the possibility that product differentiation in certain capital goods may be driven, in part, by learning by using. In many cases, the users of certain forms of capital goods (particularly instruments) may themselves make important modifications in the capital good, modifications incorporated in subsequent models. In a recent study of clinical analyzers, von Hippel and Finkelstein compare the competing models, showing that certain designs have proven more amenable to user modification and ultimately have been more successful commercially.[4]

The learning-by-using experience generates two very different kinds of useful knowledge that, borrowing from a well-established terminology, we may designate as *embodied* or *disembodied*. In the first case, the early experience with a new technology leads to better understanding of the relationship between specific design characteristics and performance that permit subsequent improvements in design. In this case, the result is an appropriate design modification. What we are describing here is a feedback loop in the development stage. Optimal design often involves many iterations. A given modification by itself may raise new, unanticipated difficulties or turn out to involve tradeoffs with other desired performance characteristics. This knowledge is pursued in prototype testing, but such testing may not disclose various kinds of useful information.

Thus, learning by using in aircraft generates highly specialized knowledge concerning the optimal design of aircraft components. This knowledge flows into a growing pool at the product development stage

[4] Eric von Hippel and Stan N. Finkelstein, "Product Designs which Encourage – or Discourage – Related Innovation by Users: An Analysis of Innovation in Automated Clinical Chemistry Analyzers," Working Paper 1011-78, July 1978.

that is also fed by other sources, such as metallurgical improvements, further progress in the miniaturization of electronic components (avionics), and so on. Out of this confluence comes a steady flow of small improvements that can be embodied in new hardware.

In the second case, that of disembodied knowledge, the knowledge generated leads to certain alterations in use that require no (or only trivial) modifications in hardware design. Here, prolonged experience with the hardware reveals information about performance and operating characteristics that, in turn, leads to new practices that increase the productivity of the hardware – either by lengthening its useful life or by reducing the operating costs. Learning by using includes both embodied and disembodied knowledge.[5]

Learning by using, in its purest form, is disembodied. As we will see, however, this process creates new information that eventually results in the physical modification of hardware. In this sense, it constitutes a feedback loop into the design aspect of new product development. Obviously, it is difficult to make a sharp distinction between the embodied and the disembodied consequences of learning by using because, in practice, the former blends into the latter. Extensive use of an aircraft may eventually lead to the discovery of faults in components or design, as in the discovery of metal fatigue that led to considerable loss of life in the Comet, or the unusual resonances that eventually weakened the engine mounts of the Electra and also led to fatal crashes, or the frequent failure under stress of the fan-jet turbine blades of the Boeing 747 in 1969–70. Such learning by using obviously required alterations in the aircraft of an embodied sort. Early service experience thus commonly leads to modifications in aircraft components. Similarly, there has often been a great deal of learning about the exact combination of aerodynamic conditions that would lead to stalling – especially but by no means exclusively in military aircraft. This information sometimes leads to hardware modification. On the other hand, other information gradually accumulates concerning optimal flight performance conditions, such as that pertaining to fuel economy. This learning may involve no hardware modifications but would be incorporated into training manuals and the instruction of flight personnel. Again in terms of disembodied knowledge, learning by using may lead to modifications in maintenance practices, such as changes in the timing of maintenance cycles. Such changes are a common occurrence.

[5] Although I am not asserting, obviously, that all embodied technological improvements are derived from learning by using.

III

The aircraft industry, in which learning by doing in production was observed in the 1940s, also provides useful evidence of both the embodied and disembodied forms of learning by using.[6] Further, in the case of embodied learning, the role of final product users (airlines) is very important in product differentiation and modification. To examine the embodied form of learning first, note that learning does not occur in the manufacturing process alone as workers improve their skills in making the product; as a result of actual use of the aircraft, learning also takes place concerning design aspects and many factors that affect the operating costs of a new model airplane. During the operation of a new aircraft, operating cost reductions depend heavily upon learning more about the performance characteristics of the system and components, and therefore upon understanding more clearly the full potential (as well as potential bugs), of a new design.[7] For example, it is only through extensive usage that detailed knowledge is gained about engine operation, maintenance needs, minimum servicing and overhaul requirements, and so on. This is due partly to an inevitable – and highly desirable – initial overcautiousness by the manufacturer in dealing with a product with which there has been no operating experience.

Much of this initial cautiousness derives from the pervasive importance of uncertainty in the prediction of performance in airplane design. In spite of elaborate experimentation in wind tunnels of increasing sophistication and theoretical techniques of increasing precision in aerodynamic research, such things as scale effects, compressibility, and turbulence continue to produce unexpected outcomes. Airflow behavior at transonic speeds has turned out to be exceedingly difficult to analyze. Sometimes performance has exceeded expectations. Sometimes there have been unexpected benefits as well as unexpected problems. Wind tunnel tests in the past, for example, resulted in exaggeration of the increase in drag, particularly at transonic speeds, and the handling

[6] For two interesting anticipations of learning by doing in aircraft production, see Adolph Rohrbach, "Economical Production of All-Metal Airplanes and Seaplanes," *The Journal of the Society of Automotive Engineers*, January 1927, and T. W. Wright, "Factors Affecting the Cost of Airplanes," *Journal of the Aeronautical Sciences*, February 1936.

[7] A parallel process, not dealt with here, is the extensive learning involved in the operation and management of an entire aircraft fleet. There were many operational problems for which optimal procedures had to be developed – scheduling problems, turnaround time, dovetailing the requirements of equipment with those of personnel, and so on. Such responsibilities belong to the realm of management and not technology, although the two are obviously interrelated.

problems associated, for example, with a swept-back wing design. With the advent of the jet engine, aircraft moved into a new high-altitude, high-speed environment where they encountered unanticipated aerodynamic effects, including lethal stalling. There was an extensive learning experience before the behavior of the new aircraft in their new environment was reasonably well understood.[8] In addition, the behavior of metals after prolonged use or with aging is still very difficult to analyze. Metal fatigue remains a nemesis in the design and construction of aircraft. Simulation methods for studying aging, methods that, for example, are supposed to accelerate the aging process of certain alloys, have not proven to be a reliable guide in the recent past.[9] The performance of new engines remains notoriously uncertain in the development process; problems must be dealt with essentially by trial and error. Thus, one must not exaggerate the extent to which, even today, the design of aircraft can draw upon precise scientific methodology.[10] Much of the essential knowledge in aircraft design and construction can still be derived only from in-flight learning.[11]

The confidence of designers in the structural integrity of a new aircraft is an increasing function of elapsed time and use. Prolonged experience with a new design reduces uncertainties concerning performance and potential, and generates increasing confidence concerning the feasibility of design changes that improve the plane's capacity. Thus, after the highly successful 727 had been in use for some years, Boeing made numerous modifications that it would not have had the confidence to make earlier. These modifications included stretching of airplane fuselages.

[8] Early 707s sometimes experienced "jet upset" – that is, they went into unaccountable dives from high-altitude flights. The XB-47 bomber encountered the disastrous "coffin corner" at certain combinations of speed and atmospheric conditions.

[9] "Steiner pointed out that 'accelerated aging' tests have not proved accurate in the past. He cited the case of certain alloys that 'aged in a most peculiar manner' a few years ago. In five to ten years, these alloys – utilized on the Boeing 707 and other transports – developed inter-granular corrosion, requiring expensive inspection procedures and replacement." "Greater Government R&D Urged to Spur Advances," *Aviation Week and Space Technology*, 12 September 1977, p. 35. Steiner was a Boeing vice-president in charge of product evaluation at the time.

[10] Ronald Miller and David Sawers, *The Technical Development of Modern Aircraft* (New York, 1968), pp. 246–50. The advent of the computer, it should be noted, has in many cases improved the predictive powers of airplane designers. Many critical uncertainties, however, persist.

[11] It is still far from unusual for engineers in many industries to develop a successful solution to a problem for which there is no scientific explanation, and for the engineering solution to *generate* the subsequent scientific research that eventually provides the explanation. For some interesting examples, see R. R. Whyte, ed., *Engineering Progress through Trouble* (The Institution of Mechanical Engineers, London, 1975).

The stretching of aircraft, so critical to the economics of the industry, has been closely tied to the growing confidence in performance generated by learning by using. Consider the succession of changes that have been embodied in the DC-8. In this aircraft, operating energy costs over its life span on a per-seat-mile basis have been reduced by 50 percent, even though the basic configuration has remained largely unchanged and the modifications have been relatively unsophisticated compared to differences between aircraft types. Clearly, an important set of modifications involves the engines. Their available thrust has been increased and their fuel consumption decreased, thus increasing the potential payload and directly reducing operating costs. At the same time, modifications to the wing profile have reduced the drag of the aircraft. With the DC-8-30, a drooped flap was added; then, a leading edge extension was designed for the DC-8-50. Subsequent models increased the aspect ratio and repositioned the flap.[12] Engine pylon design also underwent some modification. These variations in the aircraft's geometry were motivated by the drag reduction and increased fuel economy they were able to provide. These modifications played a critical role in the eventual decision to stretch the aircraft, increasing its capacity from 123 to 251 seats. It is essential to remember the interdependence of these technological improvements. The possibilities for stretching, and consequently for adding payload volume and weight to the vehicle, depended upon the availability of more powerful engines, drag reduction, and other design modifications.

The story of the DC-8 is representative of commercial aircraft industry design practices. Innovations that have been incorporated in a particular vehicle and greatly reduced its operating costs are primarily connected with (1) developing the engine to increase available thrust and reduce fuel consumption, (2) reducing drag by modifications in wing design, and (3) stretching the vehicle to increase the payload capability. Although dramatic improvements in operating costs may initially appear to come directly from the stretching process, such a gain is unattainable without the complementary developments of power plant technology and sometimes wing technology, which are themselves highly interdependent. Engine technology, particularly during the turbine era, has experienced dramatic growth in terms of thrust per pound

[12] Although the modifications alter the aerodynamic parameters of the wing, sometimes substantially, the wing itself does not generally experience internal structural alterations. This is because of the prohibitively high cost of wing redesign, which makes it much more economical to modify the flaps, leading edge, and wing tips. At the same time, the possibility of eventually utilizing even these add-on devices must be anticipated to some degree during the initial wing development stage.

of engine weight, which has increased by over 50 percent in twenty years. Even more dramatic, however, has been the record of fuel consumption per hour per pound of thrust. For example, in 1950, about 0.9 pound of fuel was required for each hour-pound of thrust. By the early 1960s, this requirement, with the development of turbofans, had dropped to about 0.75 pound of fuel per hour-pound of thrust. With the introduction of the high-bypass-ratio turbofans around 1968 and in use today, the fuel requirements dropped to 0.6 pound of fuel per hour-pound of thrust. This 30 percent decline in fuel requirements over this period has direct implications for increasing the deliverable payload of aircraft within the turbine generation.[13]

Moreover, the introduction of the high-bypass-ratio turbo engine underlines some of the subtleties of the innovation process when complex and highly interdependent technologies are involved. The essential point is that the final outcome is extremely uncertain because of the difficulty of anticipating some of the interactive consequences of any given innovation. When the high-bypass-ratio turbofan engine was introduced in the late 1960s, it was anticipated to cause a 20 percent reduction in fuel requirements per pound of thrust per hour. This expectation was fulfilled. However, this new fuel economy brought an unanticipated increase in labor costs, which consisted of sharply increased maintenance requirements – specifically, a 100 percent increase in maintenance costs per pound of thrust per hour.[14] Thus, the greater fuel economy with the new engines was attained only with a great increase in maintenance costs.

The stretching of jet transports, from the Comet to the 747, is a classic example of a process that is not very interesting technologically but is of vital economic importance.[15] This process reflects the basic complemen-

[13] Boeing Commercial Aircraft Company, Document B-7210-2-418, 5 October 1976, p. 4.

[14] "Greater Government R&D Urged to Spur Advances," p. 35.

[15] The technique of stretching has a much older history and was applied with great success to the DC6-DC7C series as well as to the Lockheed 649 to 1049H series of propeller-driven aircraft. An excellent recent example of this technique is the DC9 series aircraft, which has been increased in size by lengthening the fuselage from 104.4 feet (80 passengers) in 1965 to 147.8 feet (155 passengers) in 1980 in five distinct steps. In addition, modifications to the wing and power plant have enabled it to increase its performance and keep abreast of the latest noise regulations. See *Business Week*, 17 November 1977, pp. 95 and 100. Currently under consideration (June 1979) is yet another stretch that would lengthen the fuselage to 157 feet, 4 inches, and would expand the seating capacity to 170 passengers. See *Aviation Week & Space Technology*, June 1979, p. 24.

tarity between the performance of the engine and the airframe. Indeed, there is little incentive to improve engine design unless airframe designers know how to exploit the improvement.[16] The carrying capacity of the airplane depends, first of all, on the capacity of the engines. As engine performance is improved, exploitation of its potential requires redesign or modification of the airframe. The simplest response, as improved engines become available, is to stretch the fuselages and add more seats. Indeed, as learning by using provided a better understanding of these relationships, most airplanes were deliberately designed to facilitate subsequent stretching. Although airplane designers conform to the current capacity of the engines, it is generally understood that engines with improved and increased performance will appear within the lifetime of the model, and it is important to be able to exploit them. Since designers expect these future engine improvements (as well as other complementary technological improvements), they consciously attempt to build flexibility into the airplane. This applies especially to fuselage design to facilitate later stretching. Such stretching is important in improving productivity. Stretching may, indeed, be thought of as the process by which, as a result of accumulating knowledge and improved engine capabilities, the payload possibilities of a new airplane design are expanded to their fullest limits. The decision to undertake such stretching is, obviously, dictated by such factors as the growth of passenger demand or the availability of new routes.

[16] The role of highly specialized producers, and the question of what constitutes the optimum degree of specialization from the point of view of technological innovation, are highly important issues that are still not very well understood. Specialist producers tend to be very good at improving, refining, and modifying their product. They tend to be weak in devising the new innovation that may constitute the eventual successor to their product. They tend, in other words, to work within an established regime, but they do not usually make the innovations that establish a new regime. Thus, the buggy makers did not contribute significantly to the development of the automobile; the steam locomotive makers played no role in the introduction of the diesel, and indeed expressed total disinterest, until it was finally introduced by General Motors; and the makers of piston engines did not play a prominent role, in England, Germany, or the United States, in the development and introduction of the jet engine. The severely circumscribed technological horizons of specialized producers – to some extent an inevitable occupational hazard – may help to account for what one recent book on the aircraft industry describes as "an apparent proclivity on the part of once successful manufacturers to remain too long with the basic technology of their original success." Almarin Phillips, *Technology and Market Structure: A Study of the Aircraft Industry* (Heath Lexington Books, Lexington, Mass., 1971), p. 91. The point is that intimate familiarity with an existing technology creates a strong disposition to work within that technology, and to make further modifications leading to its improvement rather than its displacement.

IV

Let us now consider the problems associated with power plant maintenance as an example of the disembodied form of learning by using. Proper power plant operation requires extensive maintenance and, in the case of the radial (reciprocating) engines, necessitates complete overhaul after a specific time interval. In the case of the jet engine, this interval was originally based on the experience with reciprocating engines and was extended as experience was gained. Involved here was a gradual learning process; as a result, sufficient confidence in the new engine was acquired to justify some increase in the period between overhauls. The significance of this increase may be judged from the fact that, whereas the airlines had overhauled piston engines after 2,000 to 2,500 hours of service, after some years of experience with the jet engines the time interval between overhauls was as high as 8,000 hours.[17] Since jet aircraft were flying at much greater speeds than piston-driven aircraft, the amount of useful work done by the jet engines between overhauls was far greater than is indicated by the increased number of hours alone. The lengthening of the servicing and maintenance schedules, as in the case of the 707 – the early workhorse of the jet age – came only gradually as experience accumulated to strengthen confidence in the structural integrity of the aircraft.[18]

Improved maintenance of the propulsion system is important in reducing the operating cost of aircraft systems.[19] Overall maintenance typically comprises 30 percent of all direct operating costs of labor and materials. Even this figure understates the overall impact, since the lost revenues due to unscheduled maintenance requirements may be substantial as well. For example, at American Airlines, over a thirty-seven-month period between 1972 and 1975, maintenance problems accounted for an average of 21 percent of all delay costs, reaching an annual average of over $8 million in costs attributed to maintenance.[20] In terms of direct operating costs, maintenance costs for current jet vehicles are roughly equal between airframe and propulsion systems, although the activities are somewhat differentiated because airframe maintenance is more labor-intensive than engine maintenance.

[17] Miller and Sawers, *Technical Development of Modern Aviation*, p. 197.

[18] The lengthening of the time interval between maintenance checks required FAA approval.

[19] See NASA Document CR-134645, "Economic Effects of Propulsion System Technology on Existing and Future Transport Aircraft," 1974, particularly Section III.

[20] See NASA Document RFP 1-15-5595, NAS 1-14284, "The Impact of Technology on Operating Economics of Existing and Future Transport Aircraft," 1976, pp. M760816–19.

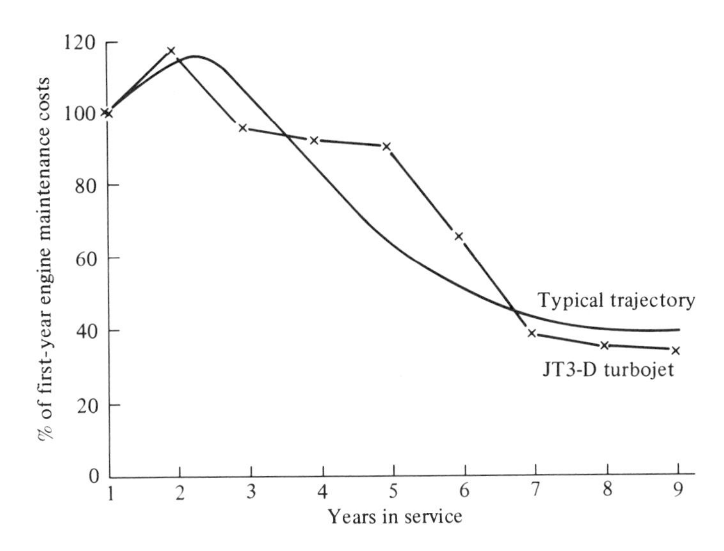

Figure 1. Engine maintenance expense. (*Source:* NASA CR-134645.)

A typical cost of engine maintenance schedules is presented in Figure 1, together with the particular trajectory for the Pratt and Whitney JT3D turbojet engine.[21] The rise of maintenance costs during the first year of introduction reflects the impact of early design problems that were not anticipated prior to the rigors of actual on-line operations. In the case of the JT3D, the design difficulties remedied through maintenance involved inordinate wearing of parts due to high operating temperatures relative to the thermal stability of the lubricants used. After this point, maintenance costs dropped sharply, typically to 30 percent of their initial levels over a decade of operation. It is the determinants of this cost reduction after the introduction of a new engine that we seek to understand.

Just as some design problems, such as the lubrication difficulties of the JT3D, are solved under the guise of maintenance, so are maintenance aspects of the future engines carefully examined and prepared for during the design phase. This activity includes preparing of instruction manuals and tools for repair, ordering and inventorying spare parts, training personnel, and so on. These are complementary, although simultaneous, technological advances in the introduction of the new en-

[21] See NASA Document CR-134645, Section II, Fig. II-1.

gine. Thus, there is an intertwining of disembodied and embodied forms of learning by using.[22]

But there is also a broad spectrum of complementary changes, with significant cost consequences, that cannot be anticipated when the propulsion system is designed. One striking aspect of this technological development is that, to a surprisingly large extent, it is not readily identified with new and innovative forms of hardware. Rather, the maintenance history of particular engines, especially turbojet engines, strongly reflects learning by using, where prior knowledge based upon reciprocating propeller engines was largely inadequate to anticipate the durability and reliability of the gas turbine engines (indeed, that earlier experience turned out to be positively misleading). This problem was further exacerbated by the fact that jet maintenance occurred first in the military, where costs were not of overriding significance. These procedures had to be modified to conform to the commercial constraints of the civilian sector.

This phenomenon is clearly reflected in the history of airline maintenance practices, in conjunction with recommendations of engine manufacturers and requirements of the Federal Aviation Authority (FAA). Resolution of these conflicting concerns has often led to substantial adjustments in maintenance. Basically, engine manufacturers are commissioned to design equipment that can be operated profitably by the airlines and that meets or surpasses the safety requirements imposed by the FAA.

These safety requirements, based upon the experience with reciprocating engines, led to systematic underestimation of reliability when they were first applied to the new gas turbines. Early jet engine maintenance programs were based upon specifications of allowable time between overhauls (TBOs) measured in hours of operating time. These were strictly enforced and were extendable only in increments of 200 hours, and then only after extensive testing of several devices. Although this practice was justified initially, due to safety considerations and ignorance of the capabilities of the new technology, these programs were extremely expensive because unnecessary maintenance work was done at excessively short intervals.

[22] The disastrous DC-10 crash in Chicago in May 1979 raises some extremely troublesome questions about the effectiveness of the intertwining in the case of this particular aircraft. If the crash was indeed caused by a faulty maintenance procedure – the failure to separate the pylon and the engine during removal and reinstallation – the aircraft designers hardly deserve complete exculpation. At the very least, the design of the aircraft apparently made the maintenance procedure extremely exacting and inherently dangerous. Surely, a basic desideratum of good design is that aircraft components should be more "forgiving" to variations in handling procedures.

When this was realized, the next stage was a modified TBO program that removed the obligatory disassembly conditional upon various tests and inspections. Today, there are no mandatory schedules; reconditioning now takes place as indicated by routine tests that are performed while the aircraft remains on-line. Such examinations include the use of borescopes to check wear, analysis of used lubricants, and visual examination. This trend can also be seen at other levels. For example, inspections of the Boeing 707, which were initially required daily, have been stretched out to routine weekly surveillance.[23]

The increased aircraft availability resulting from this improved maintenance scheduling clearly depends much more on familiarity with the hardware than on technological advances in the new hardware. However, there are clear implications for the engine hardware resulting from the learning-by-using process. Thus, engine maintenance on a need-only basis quickly pinpoints the factors that limit durability. Redesign efforts are then focused upon these elements. Further, because it is no longer necessary to recondition an engine as a complete unit, the use of interchangeable modules has resulted in reduced costs. In addition, significant technological advances have been made in the diagnostic hardware used to ascertain the advisability of maintenance. For example, more sophisticated borescopes using television transmitters for monitoring, internal accelerometers to monitor vibrations, and isotope pellets to detect metal fatigue and stress have all been introduced in the diagnostic phase of maintenance. Here again, if the role of diagnostic equipment is taken into account, embodied and disembodied forms of learning by using are intertwined.

This change from a very conservative preventive maintenance program to one based upon diagnostic techniques is reflected in the statistics on the frequency of engine removals with respect to operating times, as shown in Figure 2. Even with the new diagnostic maintenance regime, engine removals for maintenance and repair have declined over the past several years. This is shown in Figure 3, where a decline is noted for each of the first ten years. After this point, the engines reach the durability limits of major structural members, requiring increasingly frequent removal in later years.[24]

[23] Boeing Commercial Aircraft Company, Document B-7210-2-418, p. 3.

[24] There is an interesting aspect of the maintenance and replacement requirements of commercial aircraft during their normal life cycles that usually goes unnoticed. After the initial purchase of an aircraft, the replacement requirements for the power plant are far greater than those for the airframe, which helps to account for some of the intensity of recent negotiations for engine contracts in new generations of aircraft. "Rolls points out that when a customer buys an aero engine he can expect to have to spend about one and a

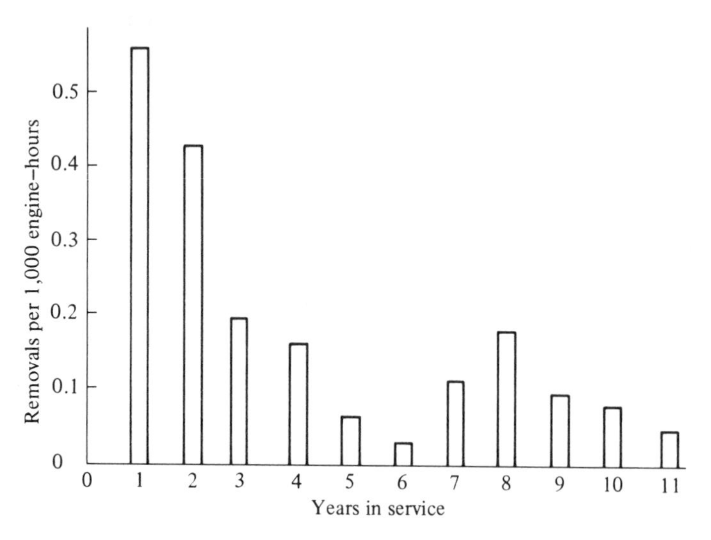

Figure 2. Engine maintenance removal rates. (*Source:* C. P. Sallee, *Economic Effects of Propulsion Systems Technology on Existing and Future Aircraft,* American Airlines, New York, 1974, p. 40.)

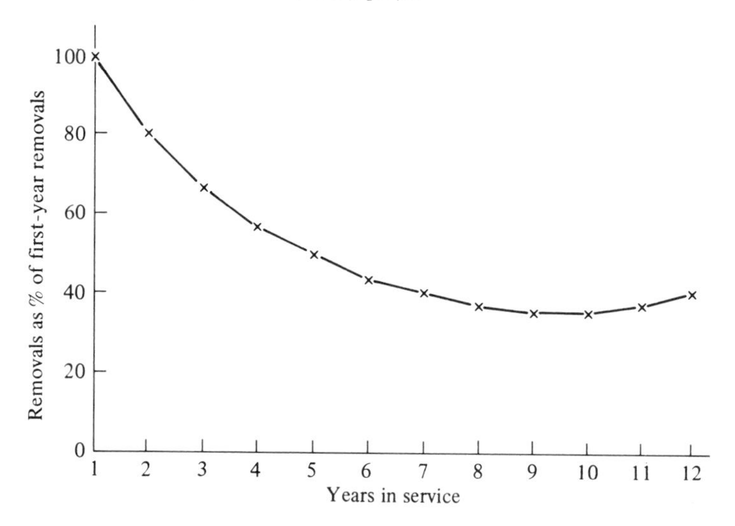

Figure 3. Engine removal for overhaul and repairs vs. time. (*Source:* C. P. Sallee, *Economic Effects of Propulsion Systems Technology on Existing and Future Aircraft,* American Airlines, New York, 1974, Figure II-3.)

Again, this pattern understates the full impact of this new maintenance schedule program, because it does not include the substantial opportunity costs no longer foregone when aircraft must be pulled out of service for frequent but unnecessary maintenance. The significantly decreased downtimes permitted by the new programs keep the aircraft in service for a higher proportion of their lifetime. The result, of course, is a significant increase in the output/capital ratio for the aircraft.

Changing propulsion system maintenance practices demonstrate the ways in which learning by using is responsible for significant productivity improvements that are gradually teased out of a major new innovation such as the turbojet engine. This source of drastically improved economic efficiency is not new; it has long characterized the adjustment of the air transport industry to new, untried equipment. Indeed, one authority on the industry has estimated that, between 1920 and 1936, engine maintenance costs of radial reciprocating engines fell by 80 percent, constituting the largest contribution of the engine manufacturer to the development of air transport at that time.[25] This contribution, however, was not and could not have been made by the engine manufacturer alone; it was possible only in conjunction with the learning by using of the air transport carriers. In fact, as should now be apparent, the active participation of the user in the process of making technological improvements is one of the critical features that distinguishes learning by using from learning by doing.

V

Even if the learning by using described here is significant for the aircraft industry, is it important elsewhere? Or, to ask a different question, what reason is there to believe that learning by using is becoming more pervasive? These questions can be answered only empirically. Let me suggest, however, that the importance of these phenomena in aircraft is an outgrowth of certain characteristics that are not unique to the aircraft industry. There is, therefore, ample a priori reason to expect learning by using to be associated with similar characteristics in other industries.

What characteristics are related to learning by using? They seem to be associated with a high degree of systemic complexity. Such complexity is the essential reason why optimal design, maintenance, and operating

half times the engine's purchase price on essential spares throughout its lifetime. This compares with only one-third the purchase price in spares for an airframe." Mark Hewish, "Airliner Numbers Game: Does It Add Up?" *New Scientist*, 31 August 1978, p. 616.

[25] Miller and Sawers, *Technical Development of Modern Aviation*, p. 89.

procedures take so long to determine. As a result, there are large fixed costs and long lead times in the development process out of which the design characteristics emerge. The products themselves have long useful lives, and there is great uncertainty about their eventual operating characteristics.[26] This is because the products are composed of many subsystems that interact in complex ways, often in ways that render highly uncertain the outcome of attempts to scale up or scale down.

Many of these same characteristics are found elsewhere and, I would conjecture, with similar learning-by-using consequences. Casual inspection, for example, suggests that electricity production in steam power plants possesses many similarities to aircraft production. Like aircraft, it is a complex system of interacting components. In addition to the high fixed costs of the equipment itself, there are huge development costs, long lead times, and considerable uncertainty about performance that can be eliminated only through use and experience. Maintenance and repair are a significant portion of operating costs, there are compelling economic reasons for minimizing downtime, and reliability is a major consideration. Thus, whereas the aircraft industry has had to deal with such problems as metal fatigue and the determination of optimal maintenance procedures, electric power has been preoccupied with fouling of various kinds – rust, corrosion, cracks, leakages, clogging, slag accumulation, ash, and so on. All of these problems have led, in the past, to high maintenance costs or a short life expectancy until improved methods of dealing with them were developed. As a result, the industry in its earlier history was plagued with frequent outages. Such outages were extremely costly because they necessitated duplication of equipment. As these difficulties were overcome, and as higher standards of reliability could be taken for granted, capital costs were considerably reduced by eliminating the need for duplication. "The high degree of operating reliability of the newer steam generating equipment makes feasible the adoption of the unit principle of one boiler per turbogenerator unit with

[26] As a result, the socially optimal amount of pretesting before a product is placed on the market, or the optimal frequency of maintenance activity, becomes extremely difficult to establish. In the case of aircraft, one could conceivably (by analogy with pharmaceutical practices) produce large fleets of new models and pretest them for many years before sanctioning their public use. Such prohibitively costly requirements would doubtless halt technical progress in commercial aircraft. In the case of maintenance procedures, there is an additional complication to the high cost of the activity itself. It is known that taking complex things apart creates some nontrivial likelihood of putting them together incorrectly. Thus, the possibility of identifying a defect by more frequent maintenance is offset by the possibility of doing something wrong and therefore of creating new hazards as well as reducing existing ones.

appreciable savings in unit investment."[27] Moreover, increasing the equipment's reliability had exactly the same consequence as stretching out the time between overhauls of aircraft jet engines.[28] Finally, even in recent years, much performance uncertainty has persisted in the operation of new, larger-scale units in the electric power industry. Such uncertainties are resolved only after extensive use and experience; the underlying problems may not even declare themselves for a few years.[29] In all these respects, learning by using phenomena in electric power generation show close similarities to those described in the aircraft industry.[30]

Reliability is a pervasive concern in high-technology industries, and the attainment of high standards of reliability is a major concern in the development stage. There are various reasons for this. In some cases, as in aircraft, reliability is absolutely essential to the service being provided.[31] This is also the case in radar installations for both military and civilian purposes.[32] In addition, insufficiently high standards of reli-

[27] *Power Stations* (Electric Power Institute, 1949), p. 19. See also p. 21.

[28] "The once-a-year scheduled shutdowns for inspection have gradually increased to once in two years with a tendency today to operate three or four years between inspections. There is reason to believe that even longer intervals are practical." Ibid., p. 29.

[29] "Even under the most favorable conditions for advancing the scale frontier, the cost side of the equation imposes fairly strict upper limits on the economical pace of advance, and trying to force the pace could mean sharply rising cost of development. The experience required for pushing out the scale frontier is related to time and cannot be acquired by increasing the number of similar new units. Perhaps the greatest uncertainties connected with units arise from problems that may not show up until the units have been in operation for a few years." William R. Hughes, "Scale Frontiers in Electric Power," in William M. Capron, ed., *Technological Change in Regulated Industries* (The Brookings Institution, Washington, D.C., 1971), p. 52.

[30] For a valuable recent attempt to characterize the nature of the learning experience in one segment of the electric power industry, see Paul L. Joskow and George A. Rozanski, "The Effects of Learning by Doing on Nuclear Plant Operating Reliability," *Review of Economics and Statistics*, May 1979.

[31] This statement requires some qualification. Commercial users of aircraft are extremely concerned with the reliability and durability of engines. The military have been less concerned with reliability and more concerned with maximum performance and peak power capabilities. Military aircraft, compared to civilian aircraft, do not fly for many hours, but their engines must function at high power levels. Civilian aircraft, by contrast, spend far more hours in flight, the engines operate at far below their maximum power (except during takeoff), and there is a powerful incentive to extend the time intervals between overhauls. As a result of these different performance criteria, there was still much to be learned about the behavior of jet engines when they were transferred from military to civilian uses in the 1950s. See Miller and Sawers, *Technical Development of Modern Aviation*, pp. 86–97.

[32] Burton Klein has commented on the preoccupation with reliability problems in the introduction of radar technology. "In radar development, probably much more has been

ability create high capital costs, as in the case of electric power generation, when backup or other duplicative facilities must be provided to deal with outages and breakdowns of essential components. Moreover, the high labor cost of maintenance and repair is a powerful motivation for reliability improvement. This issue is compounded by the fact that breakdowns are extremely difficult to deal with in industries where malfunctioning equipment is inaccessible. Classic examples include undersea cables in the communications industry – and, more recently, communications satellites in space. In fact, the desire to reduce labor-intensive maintenance and repair activities has been a major factor in the design of telephone equipment.[33] Such concerns, moreover, have been responsible for one of the most important inventions of the twentieth century – the transistor. Bell Laboratories research has for many years been geared to increasing the reliability and decreasing the maintenance costs of systems components. The research program, which was initiated by Mervin J. Kelly in the 1930s to provide a solid-state amplifier, culminated in the invention of the transistor by Bardeen, Brattain and Shockley in 1947.[34] Moreover, the systemic complexity of the telephone industry makes it virtually impossible to design equipment in the laboratory that will perform flawlessly when installed. Here, as in aircraft, ships, or electric power generation, there is an initial shakedown period when unanticipated difficulties have to be dealt with. The recent introduction of electronic switching systems created many such difficulties.[35] Such learning by using – sometimes leading to modifications

spent on overcoming reliability problems than on all other problems combined." *The Rate and Direction of Inventive Activity* (National Bureau of Economic Research, Princeton University Press, Princeton, 1962), p. 480. More generally, Klein states: "Experience indicates that one of the most unpredictable things about a new piece of equipment is the problems that will be encountered in making it perform reliably, and the time that it will take to overcome these problems. One of the main reasons why development costs tend to be seriously underestimated is a tendency to minimize the reliability problems that will be encountered and the amount of testing and modification that will be required to overcome them." Ibid., p. 482. Klein's comments would apply equally well to the construction of the Minuteman missile system, the essential problem of which was the attainment of sufficiently high levels of reliability.

[33] H. W. Bode, *Synergy: Technical Integration and Technological Innovation in the Bell System* (Bell Telephone Laboratories, Murray Hill, N.J., 1971), p. 46.

[34] L. H. Hoddeson, "The Roots of Solid-State Research at Bell Labs," *Physics Today*, March 1977, pp. 24–30.

[35] "Bell Laboratories people readily confess that in their original planning they underestimated the difficulties inherent in programming these computer-like machines to respond exactly right in every conceivable situation when people started to use them. It was only by actually putting the new system into operation and letting it cope with real traffic that all the program wrinkles could be smoothed out – a process that could never have taken place in the development laboratory." Prescott Mabon, *Mission Communications. The Story of Bell Laboratories* (Bell Telephone Laboratories, Murray Hill, N.J., 1975), p. 7.

in product design, sometimes to alterations in maintenance practices or merely to alterations in operating procedures – appears to be characteristic of more and more high-technology industries.[36]

The creative use of learning by using as a business strategy may now be an important factor in some high-technology industries. Consider the computer industry, which, in recent years, has relied increasingly on complex software products to make its systems useful to a broader range of users. The development of effective software is highly dependent upon user experience. The modification of software systems in response to this experience is now intrinsic to software engineering. This is so because most software products permit wide variations in inputs and processing options. These options cannot possibly be tested completely prior to the release of software. Thus, the optimal design of software depends upon a flow of information from its customers. Furthermore, many computer companies routinely provide extensive software support that involves software modification when bugs are discovered by customers – as they inevitably are – when the software is used. The effectiveness of support services in improving the product after its release appears to be very important to the competitive success of computer firms. Such service arrangements represent, in effect, an institutionalization of procedures for exploiting the learning by using phenomenon in the computer industry.

In software engineering, there is a related aspect of learning that involves doing rather than using, as these terms are employed here. In order to develop efficient, large software systems consisting of many separate modules, some sections must be rigorously tested to determine basic characteristics, such as speed and reliability, before other modules can be completed. This sequential design process, in which each step is dependent upon information generated in the previous step, appears to be essential to the design of large software systems and may be a basic principle in other large system design problems.[37]

The increasing reliance upon modular systems is, at the same time, altering the meaning of maintenance activity. "Because minicomputers are designed modularly on plug-in LSI (large-scale integration) circuit boards, and repairs to the board are usually done only at the factory,

[36] The dependence upon vacuum tubes in the logic circuits of early computers involved very serious problems of reliability, with resulting high costs of replacement and maintenance – as is immediately apparent from the fact that ENIAC contained no less than 18,000 tubes. As a result, when the transistor became available after its development at Bell Labs, its much greater reliability as compared to the vacuum tube was one of the compelling reasons for its rapid introduction into the computer industry.

[37] See Frederick Brooks, Jr., *The Mythical Man-Month* (Addison-Wesley, Reading, Mass., 1975), especially chapter 2, for some illuminating comments on the nature and consequences of sequential constraints in computer programming.

maintenance consists in many cases of plugging in a new board. Users can stock spare boards and plug in a new one themselves."[38] The decreasing costs of electronic circuitry, as compared to the high labor cost of maintenance and servicing, will doubtless lead to further extensions of modular design.

Thus, I suggest that further empirical examination of learning-by-using phenomena should be rewarding. It promises to improve our understanding of technological change by identifying more clearly the nature and locus of activities that generate such change. I suggest that learning by using is relevant to a large and growing percentage of capital equipment, especially components characterized by a high degree of systemic complexity. Secondly, the improvements included in learning by using play an important role in the decision to adopt new technologies. Thus, a clearer identification of the activities discussed here may unravel some of the secrets of the time path of the diffusion process. For example an intuitive familiarity with learning by using, and the time that must often elapse before performance uncertainties are resolved, may constitute an important reason for the decision of private firms to postpone the adoption of an innovation.[39] Finally, and as a consequence, a better understanding of learning by using may further clarify the ways advanced industrial societies generate productivity improvements and, therefore, economic growth. Valuable information of great economic relevance is, as I have argued, an important and slowly emerging by-product of the use of new technologies. This particular learning experience may therefore hold certain keys to productivity improvement in high-technology industries. For it is probably no accident that the industries that rely heavily on learning by using – aircraft, electric power generation, telephones and, more recently, computers – have had some of the most impressive productivity growth in the twentieth century.[40]

[38] Auerbach Publishers, Inc., *Auerbach on Minicomputers* (Petrocelli Books, New York, 1974), p. 15. In fact, the term *maintenance* has taken on a new and unique meaning in the programming context of the data processing industry – a meaning that flows directly from learning by using phenomena. "Maintenance is concerned with two distinct aspects of software: one is the correction of remaining flaws and errors; the other is the modification and improvement that results from changes in the environment and from the user's desires, requests and demands, or from the programmer's perception and anticipation of them." T. A. Dolotta et al., *Data Processing in 1980–1985* (Wiley, New York, 1976), pp. 94–5.
[39] For some related considerations, see Nathan Rosenberg, "On Technological Expectations," *Economic Journal*, September 1976.
[40] Productivity growth in electric power generation began a distinct decline compared to its earlier growth performance in the 1960s. To what extent this decline can be attributed to the phenomena with which this paper is concerned, as opposed to the exhaustion of scale economies, the impact of environmental regulations, or a significant decrease in innovation in the industries that supply electric power with inputs is far from clear.

7 How exogenous is science?

I

I begin by recalling Kuznets's view that the distinctive feature of modern industrial societies is their success in applying systematized knowledge to the economic sphere, knowledge derived from scientific research.[1] This view has the disconcerting aspect, at least for the economist, of appearing to make the central feature of modern economic growth an exogenous phenomenon. If that is really the way the world is, then we ought to acknowledge it gracefully and accept the fact that the major determinants of a central economic phenomenon lie outside the economists' range of analysis. Economists have many reasons for humility these days, and perhaps one additional reason will not be too marginally burdensome.

On the other hand, perhaps we do not need to be quite so humble. Perhaps if we do not insist on too restrictive a definition of the subject matter of our discipline, it may be possible to identify significant chains of causation running from economic life to science as well as from science to economic life. This is in fact what I propose to do. My argument turns, first, upon admitting technology into the arena of economic variables. Economists have had much more success in dealing with the *consequences* of technological change than with its determinants. Nevertheless, the extensive labors of Schmookler, Griliches, Mansfield and others provide sufficient warrant for admitting technology into the arena of phenomena about which economists have intelligent – and perhaps even useful – things to say.

The central theme I wish to develop is that technological concerns shape the scientific enterprise in various ways. I believe we can learn a

This paper was presented at a conference held at Harvard University in April 1981 in celebration of Simon Kuznets's eightieth birthday. It draws, in part, upon research at Bell Laboratories in Murray Hill, New Jersey. The present formulation has benefited from the incisive comments of Moses Abramovitz.

[1] See, for example, Simon Kuznets, *Modern Economic Growth* (Yale University Press, New Haven, Conn., 1966), chap. 1.

great deal about the activities of scientists – even those of scientists engaged in basic research – by starting our inquiry in the realm of technology.

It is, of course, easy enough to say that science is not entirely exogenous. The difficult issue is to attempt to specify the linkages between economics and science. Can we get beyond the level of mere platitudes? I think we can, in part because research has become very costly in the twentieth century. The willingness of society to provide financial support may have mattered less in, say, the eighteenth century, when the needs of research were likely to be fairly modest – a kite, some string, a jar, an electric storm, and a lot of good luck. It matters much more at the present time, when "little science" has been replaced by "big science" and the table stakes for conducting research may be access to a linear accelerator costing tens of millions of dollars.

This paper, then, is a kind of preliminary reconnaissance, the beginning of an attempt to develop a conceptual framework that will improve our understanding of the connections between science and economic performance. In view of the obvious and compelling importance of this subject, I offer only a token apology for the fact that this paper is, at best, only the first small step on a long intellectual journey. I will argue that technology influences scientific activity in numerous and pervasive ways. I will attempt to identify some of the most important categories of influence and to sharpen our understanding of the causal mechanisms at work.

Of course, the influence of certain technological concerns on the growth of scientific knowledge has long been recognized. Torricelli's demonstration of the weight of air in the atmosphere, a scientific breakthrough of fundamental importance, was an outgrowth of his attempt to design an improved pump.[2] Sadi Carnot's remarkable accomplishment in creating the science of thermodynamics was an outgrowth of the attempt, a half century or so after Watt's great innovation, to understand what determined the efficiency of steam engines.[3] Joule's discovery of the law of the conservation of energy grew out of an interest in alternative sources of power generation at his father's brewery.[4] Pasteur's development of the science of bacteriology emerged from his attempt to deal with problems of fermentation and putrefaction in the

[2] I. B. Cohen, *Science: Servant of Man* (Little, Brown and Co., Boston, 1948), pp. 68–71. The scientific finding, in turn, led immediately to a new scientific instrument, the barometer.

[3] D. S. L. Cardwell, *From Watt to Clausius* (Cornell University Press, Ithaca, N.Y., 1971).

[4] J. G. Crowther, *Men of Science* (W.W. Norton and Co., New York, 1936), chap. III.

French wine industry. In all these cases, scientific knowledge of a wide generality grew out of a particular problem in a narrow context. Such a recitation, however, gives only a very limited sense of the nature and extent of the interplay between science and technology. Indeed, that sense is totally suppressed in the prevailing formulation of our time, in which it is common to look at causality as running exclusively from science to technology, and in which it is common to think of technology as if it were reducible to the application of prior scientific knowledge. Thus, it seems to be quite worthwhile to examine the science–technology interaction with greater care.

II

Let me now be more specific. One of the more misleading consequences of thinking about technology as the mere *application* of prior scientific knowledge is that this perspective obscures a very elemental point: Technology is itself a body of knowledge about certain classes of events and activities. It is not merely the application of knowledge brought from another sphere. It is a knowledge of techniques, methods, and designs that work, and that work in certain ways and with certain consequences, even when one cannot explain exactly why. It is therefore, if one prefers to put it that way, not a fundamental kind of knowledge, but rather a form of knowledge that has generated a certain rate of economic progress for thousands of years. Indeed, if the human race had been confined to technologies that were understood in a scientific sense, it would have passed from the scene long ago.

Technological knowledge was long acquired and accumulated in crude empirical ways, with no reliance upon science. Scientific knowledge, of course, would have vastly accelerated the acquisition of such knowledge, but historically, vast amounts of technological knowledge were collected and exploited, and this trend continues today. So long as sufficiently powerful incentives have existed, knowledge covering a wide range of productive activities was accumulated – if only slowly and painfully. Even today, much productive activity is conducted without a deep scientific knowledge of why things perform the way they do. We operate blast furnaces even though we do not understand their combustion process very well, and we routinely fly in airplanes the optimal designs of which are achieved by fairly ad hoc, trial-and-error processes because there are no theories of turbulence or compressibility adequate to determine optimal configurations in advance. Extensive testing and modification based upon test results are still required. This is a major reason for the enormous development costs of modern aircraft.

Thus, the normal situation in the past, and to a considerable degree also in the present, is that technological knowledge has *preceded* scientific knowledge. Given the economic incentives underlying technological innovation, it should hardly be surprising that technological improvements based upon technological knowledge alone commonly take place in *advance* of scientific understanding. Commercial success demands something that will work, subject to various criteria imposed by the producer and the user. Product design engineers are involved in a complex optimizing procedure, but one in which success can be attained without scientific understanding of the phenomena involved. The *lack* of scientific understanding need not be, and fortunately often is not, an insuperable obstacle. Thus, it is to be expected that workable technological knowledge is likely to be attained before the deeper level of scientific understanding. At least this is so if sufficiently powerful economic incentives are at work.

As a result, technology has served as an enormous repository of empirical knowledge to be scrutinized and evaluated by the scientist. It is still far from unusual for engineers in many industries to solve problems for which there is no scientific explanation, and for the engineering solution to *generate* the subsequent scientific research that eventually provides the explanation.[5] This sequence has, of course, been less common in industries *founded* on scientific research – for example, those based on electricity. But even here, practical experience with the new technology often precedes scientific knowledge – by providing the unexpected observation or experience that gives rise to fundamental research. For example, in the early days of radio, amateurs were assigned the range of short-wave signals – less than 200 meters – precisely because the authorities thought that nothing much could be done with such waves. As it turned out, clever amateurs, who did not *know* that nothing could be done, quickly demonstrated that effective transmission was possible in the short-wave range. Establishing precisely why this performance so vastly exceeded expectations led to major discoveries on the nature of the ionosphere.[6] In the late 1940s, engineers looking for the cause of short circuits in electronic equipment found that they were caused by a filamentary growth – so-called whisker crystals. The discovery that these crystals were both strong and flexible led to extensive research on the conditions determining their growth and physical properties. This research led eventually to an understanding of the

[5] For some interesting examples, see R. R. Whyte, ed., *Engineering Progress through Trouble* (The Institution of Mechanical Engineers, London, 1975).

[6] Cohen, *Science*, chap. 16.

fundamental science of crystal growth, which, in turn, was of great value to the electronics industry.[7]

With respect to metals, the science of metallurgy really began to develop in the second half of the nineteenth century. The purpose was to account for the behavior of metals already being produced by the Bessemer and post-Bessemer technologies. A particularly fruitful field of scientific research lay in trying to account for specific properties of steel produced by certain technologies or exploiting particular inputs. Such phenomena as deterioration with age or the brittleness of metals made with a particular fuel were intriguing to scientifically-trained people. Nevertheless, new alloys continued to be developed even in the twentieth century by trial-and-error methods.

A new, superior aluminum alloy such as Duralumin was more or less accidentally developed and used for years before anyone really understood the phenomenon of "age hardening." That came later, and only with the introduction of instrumentation, including x-ray diffraction techniques and the electron microscope.

Wilm discovered age hardening, a finding of major significance in metallurgy, as a result of apparent inconsistencies in hardness measurements involving some aluminum alloy specimens. At the time, it was impossible to connect age hardening with structural changes that could be observed under a microscope, and no satisfactory explanation of the phenomenon arose.[8] Nevertheless, Duralumin was widely used in the aircraft industry (including the construction of Zeppelins) during the first World War. Age-hardened and precipitation-hardened[9] alloys were utilized in a growing number of commercial applications during the interwar years, and the obvious commercial value of new and superior alloys gave a powerful impetus to fundamental research that would link performance characteristics of the alloys with their underlying crystallography and atomic structure. Nevertheless, it was the practical metallurgists who made available to the engineers an array of new metals of

[7] "Mechanism of Crystal Growth Discovered," *Bell Laboratories Record*, April 1964, p. 142.

[8] "Although he must be considered a successful applied researcher and careful experimentalist even by modern standards, Wilm did not express any deep curiosity concerning the reasons for the hardening and preferred to leave to others even the speculation concerning its nature." H. Y. Hunsicker and H. C. Stumpf, "History of Precipitation Hardening," in Cyril Stanley Smith, ed., *The Sorby Centennial Symposium on the History of Metallurgy* (Gordon and Breach Science Publishers, New York, 1965), p. 279.

[9] An alloy that ages at room temperature is considered an age-hardening alloy, whereas one that requires precipitation at a higher temperature is classed as a precipitation-hardening alloy. W. Alexander and A. Street, *Metals in the Service of Man* (Penguin Books, Harmondsworth, 1954), p. 176.

much greater strength, superior strength/weight ratios, strength–conductivity combinations, and magnetic properties many years before their performance could be explained at a deeper level. Indeed, the determination to account for the performance characteristics discovered by the metallurgists and already incorporated into numerous industrial practices was a major incentive to fundamental research. "From these studies have arisen a better insight into deformation and strengthening mechanisms, additional support for dislocation and magnetic theories and verification of the existence and importance of lattice vacancies."[10]

I suggest that, even well into the twentieth century, metallurgy can be characterized as a sector in which the technologist typically "got there first," i.e., developed powerful new technologies in *advance* of systemized guidance by science. The scientist was confronted by the technologist with certain properties or performance characteristics that demanded a scientific explanation. Such technological breakthroughs as Taylor and White's development of high-speed steel (1898) and the development, in the 1920s, of sintered tungsten carbide are classic instances of technological improvements preceding and giving rise to scientific research. Indeed, Frederick Taylor had been concerned with questions of shop management and organization, and was not even familiar with the rudimentary metallurgical knowledge available to the technologists of his own time.[11]

In petroleum refining, similarly, improvements in the refining process involved the use of fractional distillation techniques that had already been developed – in this case, in other industries. Cracking methods were introduced by people with extensive practical experience but no formal training in chemistry. When Berthelot, the French chemist, "published in 1867 his basic researches into the action of heat on various hydrocarbons, he merely provided a basis for interpreting what was happening in practice in the oil industry."[12]

This sequence, by which technological knowledge preceded scientific knowledge, has by no means been eliminated in the twentieth century. Much of the work of the scientist today involves systematizing and restructuring the knowledge and the workable, practical solutions and methods previously accumulated by the technologist. Technology has

[10] Smith, *Sorby Centennial*, p. 309. See also Hugh O'Neill, "The Development and Use of Hardness Tests in Metallographic Research," ibid.

[11] See, for example, Melvin Kranzberg and Cyril Stanley Smith, "Materials in History and Society," Part I in Morris Cohen, ed., *Materials Science and Engineering* (Elsevier, Amsterdam, n.d.).

[12] Kendall Birr, "Science in American Industry," in D. Van Tassel and Michael G. Hall, eds., *Science and Society in the U.S.* (Dorsey Press, Homewood, Ill., 1966), pp. 60–1.

shaped science in important ways because it acquired some forms of knowledge first and provided data that, in turn, became the explicanda of scientists, who attempted to explain or codify them at a deeper level.

A very similar process seems to have occurred in one of the most notable accomplishments of the twentieth century – the transistor. There had been extensive prior empirical experience with semiconductor materials before Mervin Kelly, executive vice-president of Bell Labs, decided to support the research project that led, in 1947, to the discovery of the transistor effect. The peculiar behavior of semiconductors, we now know, is a function of the presence of mobile (conducting) electrons, which, in turn, is dependent upon such things as impurities, light, heat, or electrical stimuli. Nevertheless, long before these phenomena were understood, extensive use was being made of semiconductors such as copper oxide and silicon rectifiers, which had been discovered by purely empirical means. Indeed, the evidence of these already-working technologies was critical in the decision to undertake the solid-state research that culminated in the discovery of the transistor effect.[13]

III

Thus far, I have asserted that the accumulation of technological knowledge has provided a base of observations that eventually became grist for the scientific mill. Let me now add a different dimension to this interaction. In considering the impact of technology upon science, a central theme of my interpretation is that technological progress plays a very important role in formulating the subsequent agenda for science. The natural trajectory of certain technological improvements identifies and defines the limits of further improvement, which, in turn, focuses subsequent scientific research. In the aircraft industry, for example, improved performance continually brought the technology to performance *ceilings* that could be exceeded only by understanding better certain aspects of the physical world. The introduction of the turbojet had a profound impact upon science, as well as upon the aircraft industry, by progressively pushing upon the limits of scientific understanding

[13] J. A. Morton, *Organizing for Innovation* (McGraw-Hill Book Company, New York, 1971), pp. 46–8. The phenomenon of rectification was particularly intriguing because of the great discrepancy between theoretical predictions and the amount of rectification that could be observed or achieved experimentally. See G. L. Pearson and W. H. Brattain, "History of Semiconductor Research," *Bell Telephone System Technical Publications*, Monograph 2538 (Murray Hill, N.J., 1955), pp. 4–7. Much valuable and relevant experience had also been acquired as a result of the research on microwave frequencies that had been stimulated by interest in radar during the Second World War.

and by identifying areas requiring further research before additional technological improvements could occur. Thus, the turbojet first led to the creation of a new supersonic aerodynamics,

> only to give way to aerothermodynamics as increasingly powerful turbojets pushed aircraft to speeds at which the generation of heat on the surface of the aircraft became a major factor in airflow behavior. Eventually, turbojet-powered aircraft would reach speeds at which magnetothermodynamic considerations would become paramount: temperatures would become so great that air would dissociate into charged submolecular ions.[14]

Technological improvement does more than generate the need for specific types of new knowledge. The advance in knowledge frequently occurs only by actual experience with a new technology in its operating environment. Although wind tunnel testing has been an invaluable source of information on the performance of a new type of aircraft, there has always been a sizeable margin of error between these tests and actual performance. In part, this is due to an inadequate scientific theory for relating experimental test conditions to real-life situations. In part, it is due to inadequate information or the inability to conduct valid experiments. Thus, one concern over the first flight of the space shuttle was that the reentry of a winged vehicle into the earth's atmosphere involved aerodynamic considerations that were not fully understood.

One of the central features of high-technology industries, I suggest, is precisely this pattern: Technological progress identifies, in reasonably unambiguous ways, the directions of new scientific research offering a high potential payoff. This pattern may take a variety of forms. In the case of the jet engine, functioning at increasingly high speeds, the technology pointed to specific natural phenomena in a particular environment. In the telephone industry, transmission over longer distances, and the introduction of new modes of transmission, have generated much basic research. In order to improve radio transmission, it was essential to understand more clearly how electromagnetic radiation interacts with various atmospheric conditions. Indeed, some of the most basic scientific research projects of the twentieth century have grown directly out of the attempt to improve the quality of sound transmission by telephone. Dealing with various kinds of interference, distortion, or

[14] Edward W. Constant II, *The Origins of the Turbojet Revolution* (Johns Hopkins University Press, Baltimore, 1980), p. 240; and Theodore von Karman, *Aerodynamics* (Cornell University Press, Ithaca, N.Y., 1954). The growth of the railroad system played a very similar role in the nineteenth century, especially in connection with metal fatigue and strength of materials.

attenuation of electromagnetic signals that transmit sound has profoundly enlarged our understanding of the universe.

Two fundamental scientific breakthroughs, fifty years ago by Jansky and more recently by Penzias and Wilson, occurred as a result of attempts to improve telephone transmission. This involved dealing with sources of noise. In both cases, it is also worth noting, the scientific breakthrough involved the use of extremely sensitive equipment that had been developed at Bell Labs for research projects at a more applied level. Jansky was using a rotatable antenna designed by Harald Friis for dealing with problems connected with weak radio signals. Penzias and Wilson were using a remarkably sensitive horn antenna that had been built for the Echo and Telstar satellite communication projects.[15]

Jansky had been asked to deal with the problems of radio static after the creation of the overseas radiotelephone service in the late 1920s. In 1932 he published a paper identifying three sources of noise: local thunderstorms, more distant thunderstorms, and a third source, which Jansky identified as "a steady hiss static, the origin of which is not known." It was this *star noise,* as it was called, that marked the birth of radio astronomy.

Jansky's experience underlines one of the reasons why it is so difficult to distinguish between basic and applied research. Fundamental breakthroughs often occur while dealing with applied or practical concerns. Attempting to draw that line on the basis of the motives of the person performing the research – whether there is a concern with acquiring useful information (applied) as opposed to a purely disinterested search for new knowledge (basic) – is, in my opinion, a hopeless quest. Whatever the ex ante intentions in undertaking research, the kind of knowledge actually acquired is highly unpredictable. Historically, some of the most fundamental scientific breakthroughs have come from people, such as Jansky, who thought they were doing applied research.

Bell Lab's support of basic research in astrophysics is related to the problems and possibilities of microwave transmission, especially the use of communication satellites for such purposes.[16] Penzias and Wilson

[15] In an interview after receiving the Nobel Prize, Wilson said that it was the availability of that antenna that had first motivated him to work at Bell Labs. "The thing that originally attracted me to Bell Labs was the availability of the horn reflector – built for Project Echo – and traveling-wave maser, which made a unique radio telescope." Steve Aaronson, "The Light of Creation – an Interview with Arno A. Penzias and Robert C. Wilson," *Bell Laboratories Record,* January 1979, p. 13.

[16] At very high frequencies, rain and other atmospheric conditions become major sources of interference in transmission. This form of signal loss has been a continuing concern in the development of satellite communication. It has led to a good deal of research at both the technological and basic science levels – for example, the study of polarization pheno-

first observed the cosmic background radiation, which is now taken as confirmation of the "big bang" theory of the formation of the universe, while attempting to identify and measure the sources of noise in their receiving system and in the atmosphere. They found that "the radiation is distributed isotropically in space and its spectrum is that of a black body at a temperature of 3 degrees Kelvin."[17] Although Penzias and Wilson did not know it at the time, the characteristics of this background radiation were just what had been predicted earlier by cosmologists favoring the big-bang theory.

Notice that, in the telephone industry, practical goals related to sound transmission have led to highly creative basic research. But notice also that dealing with problems of communication has led to very different *kinds* of basic research as well. The outcome of the scientific research may be an intellectual tool or concept, or a new framework for thinking that may be applicable to a wide range of phenomena in very different disciplines. Thus, although Shannon's information theory[18] had major implications for the design of new equipment and new systems in the telephone industry, it also had great applicability elsewhere. Shannon offered a generalization for calculating the maximum capacity of a communication system for transmitting error-free information – an intellectual achievement that was clearly relevant to the telephone industry, where a precise understanding of channel capacity is central to engineering design. As is often the case, however, a breakthrough in one area had an impact in remote places. For Shannon's central notion, that it is possible to give a quantitative expression to information content, had numerous ramifications. Information theory is really a new way of thinking about a range of problems that occur in many places, and it has powerfully influenced the design of both hardware and software. The research output in this case has been a family of mathematical models of wide generality, with applications in the behavioral sciences as well as the physical sciences and engineering.[19]

mena. See Neil F. Dinn, "Preparing for Future Satellite Systems," *Bell Laboratories Record,* October 1977, pp. 236–42.

[17] *Impact,* prepared by members of the technical staff, Bell Telephone Laboratories, and the Western Electric Company Patent Licensing Division, M. D. Fagen, ed. (Bell Telephone Laboratories, Murray Hill, N.J., 1972), p. 87.

[18] Claude Shannon, "A Mathematical Theory of Communication," *Bell System Technical Journal,* July 1948.

[19] This point, regarding the transfer of techniques, concepts, and research tools and methods from one scientific discipline to another, is an extremely important one that has not, so far as I know, been studied in a very systematic way. The ability to transfer either hardware or software from one discipline to another has been a powerful force in the growth of research in particular fields. As Harvey Brooks has pointed out: "Molecular

There is, I am suggesting, a compelling internal logic to certain industries – for example, the telephone system – that propels the research enterprise in specific directions. Consider some of the materials needs of the system. In the development of the transistor, standards of materials purity had to be attained that were unprecedented for industrial purposes. Because transistor action was dependent on introducing a few foreign atoms on the semiconducting crystal, remarkably high standards of semiconductor purity had to be attained. To introduce a single foreign atom for each 100,000,000 germanium atoms meant that the telephone system had to attain levels of purity that presupposed a good deal of materials research.

The growth of the telephone system also meant that equipment and components had to perform under extreme environmental conditions, from transatlantic cables to geosynchronous satellites. These environmental extremes have one particularly important consequence: severe economic penalties for *failing* to establish very high standards of reliability. There are compelling reasons to attain high standards that are absent in, say, consumer electronics, not to mention a brick factory. The failure of a submarine cable placed on the ocean floor involves huge repair and replacement costs in addition to a protracted loss of revenue. Similarly, communication satellites *had* to be remarkably reliable and strong simply to survive the act of being launched and placed in orbit. They had to survive extremes of shock, vibration, temperature range, radiation, and so on.[20] Thus, high standards of reliability are not a marginal consideration but the essence of successful economic performance in this industry. This consideration has had a great deal to do with the high priority of materials research at Bell Labs over the past several decades. It is also basic to research in other directions. One of

biology was made possible in part by the application of the tools and experimental techniques of physics and was partly created by converted physicists. The surge of interest in the earth sciences – solid-earth geophysics, atmospheric physics, and physical oceanography – has been partly created by the application of physics techniques and concepts in these fields, which have made it possible to ask and answer types of scientific questions that were completely beyond the scope of observation a few years ago. All these sciences have changed from a purely observational and descriptive mode and style to a mode in which laboratory experiments and testable mathematical models are important techniques" ("Physics and the Polity," *Science*, 26 April 1968, p. 398). The reasons for these transfers, their timing, and the circumstances that are conducive to them are not very well understood. Nevertheless, such transfers appear to be an important determinant of the direction of scientific progress in the twentieth century, and it is worth pointing out that, at an earlier date, the physical sciences were the considerable beneficiaries of instrumentation development in biology and medicine – especially x-ray technology and both optical and electron microscopy.

[20] See Morton, *Organizing for Innovation*, pp. 24–5.

the main problems with the vacuum tube was its unreliability. The decision to undertake a basic research program in solid-state physics, which culminated in the development of the transistor, was strongly influenced by this and other sources of dissatisfaction.[21] But the transistor, although it eventually became highly reliable, also suffered from reliability problems in its early years. In the early 1950s, as the transistor experienced a widening range of applications, serious reliability defects emerged. These defects were linked to certain surface phenomena. As a result, major research was undertaken on the basic science of surface states. This research eventually solved the reliability problem and generated a great deal of new fundamental knowledge regarding surface physics.

But much earlier – in the 1920s – dissatisfaction with vacuum tubes had been responsible for a beautiful piece of fundamental research. C. J. Davisson had been concerned with the performance of vacuum tubes and with the possibilities for their improvement – longer life, extended capabilities, greater economy of design, and so on. He was struck by the patterns of emission from crystals of nickel observed while conducting research on "secondary emission" of electrons in thermionic tubes. Davisson demonstrated the wave nature of matter experimentally by bombarding a nickel crystal with a stream of electrons and studying the manner of their diffraction. Davisson shared the Nobel Prize in 1937 for his authoritative demonstration of the wave nature of matter.[22]

IV

There are other powerful reasons why the relations between science and technology cannot be adequately described by visualizing scientific research as appearing first, eventually leading to applications in technology. Many aspects of a material are not explored scientifically until the material has been used for a long time. This is because many problems connected with the use of a new material take time to emerge. A major concern of materials research has been to improve performance by eliminating problems that often emerge only after prolonged use. Many materials are subject to an all-too-familiar and depressing litany of degradation, fracturing, contamination, aging, corrosion, brittleness under complex stress, and a host of related maintenance difficulties.

[21] Ibid., pp. 46–8, and J. A. Morton, unpublished interview at Murray Hill, N.J., 29 November 1962, pp. 21–5.

[22] Prescott Mabon, *Mission Communications* (Bell Telephone Laboratories, Murray Hill, N.J., 1975), p. 97.

Thus, a great deal of research was conducted at Bell Labs on polyethylene before its widespread use on cable sheathing and wire insulation. Nevertheless, a whole new generation of problems arose *after* it had been installed. Much additional research, stimulated in part by these on-line difficulties, led to a much deeper understanding of its solidification pattern, or morphology. Out of this second generation of research came a much deeper understanding of how this morphology determines important mechanical, electrical, and chemical properties.[23] The brittleness of polyethylene turned out to be influenced by its molecular weight and crystallinity, and so its molecular weight was increased. The disturbing tendency of polyethylene to oxidize readily was counteracted by the development and use of antioxidant compounds, and so on.

The growth of knowledge is much more cumulative and interactive than is realized, especially when it is thought of as a one-shot, once-and-for-all affair, with new scientific knowledge supposedly leading to technological applications – period. In fact, continuing experiences with a material in a new environment, subject to new stresses, throw up new problems not dealt with, or even anticipated, before.

High-technology industries, by pushing against the limits of technical performance, are continually identifying new problems that can be addressed by science. At the same time, the prospective improvement in performance or reduction in costs promises large financial rewards. The intriguing question, of course, is why this mechanism seems to work so much better in some industries – or some firms – than others.

[23] "An Interview with Dr. Bruce Hannay," *Bell Laboratories Record,* February 1969, pp. 45–52. As W. O. Baker has pointed out, "the aspect of polymer science that has enabled so dramatic an expansion into telecommunications . . . within the past two decades, is the remarkable information transfer between behavior of the chemical entity (that is, the single average polymer molecule) and its physical embodiment, as in viscoelastic fluid form, or in solids, and ultimately in the increasingly adapted crystal itself. This is where polymer science is favored both intellectually and materially, for in the case of so many classes of matter, the coupling of bulk properties (tangible strength, resilience, friction, electrical nature, and so forth) with the basic structural unit, the atom or molecule, is far weaker. Metals are the classic other extreme, where the solid and liquid properties are almost wholly dominated by the aggregate although, of course, the total electronic structure of the single atom is of central significance. In addition, however, the individual molecular units in polymers themselves operate on a time scale of both mechanical and electrical relaxation such that a great range of conformations and of temporal responses can be obtained in their application. These factors are reflected in the growth of amounts of polymers used in the past decade by the Bell Telephone System." "The Use of Polymer Science in Telecommunications," *Annals of the New York Academy of Sciences,* p. 620.

V

I have been arguing that the scientific research agenda is closely linked to the ongoing technological needs of industry. I have also been suggesting specific avenues of such influence that can help us to understand the changing directions of scientific research. In the case of the telephone industry, specific problems and directions of research have been suggested by expansion of the network, transmission over greater and greater distances, associated problems of interference and distortion, the need to establish very high standards of reliability, and so on. The growth of the telephone system has brought it up against some other elemental constraints, one of the most basic of which is channel capacity. Historically, one of the most important and ingenious avenues of research has involved increasing the message-carrying capacity of existing channels. Eventually, however, the need to develop new channel capacity became unavoidable. Thus, there has been a movement from wires, to coaxial cables, to microwave radio systems, to satellites, and to glass fibers and lightwave transmission. In the case of lightwave communication, its attractiveness and the decision to attempt to develop it as a new mode of transmission were influenced by another increasingly binding constraint: physical space. The sheer congestion of limited conduit space in metropolitan areas gave a great impetus to research on glass fibers, which held out the prospect of cramming far greater channel capacity into existing space.

The development of optical fibers is particularly interesting because it illustrates several of the dynamic interactions with which I am concerned. Although its attractiveness as a new mode of transmission was increased by space and congestion constraints, its feasibility was rooted in another set of technological breakthroughs of the 1950s. It was the development of laser technology that made it possible to use optical fibers for transmission. This possibility, in turn, pointed to the field of optics, where advances in knowledge could be expected to have high potential payoffs. As a result, optics as a field of scientific research has had a great resurgence in recent years. It was converted by changed expectations, based upon past and prospective technological innovations, from a relatively quiet intellectual backwater to a burgeoning field of scientific research. The causes were not internal to the field of optics but were based upon a radically altered assessment of new technological possibilities – which, in turn, had their roots in the earlier technological breakthrough of the laser.

The relationship that I am suggesting here – of changes in the technological realm giving rise to fundamental research – is hardly unique to optics. Solid-state physics, presently the largest subdiscipline of physics, attracted only a few physicists before the advent of the transistor. In fact, the subject was not even taught at most universities. The training in solid-state physics that Shockley received at MIT in the 1930s was probably unavailable at any other university in America at the time, with the exception of Princeton. This situation was transformed, of course, by the invention of the transistor in 1948. The transistor demonstrated the potentially high payoff of solid-state research and led to a huge concentration of resources in that field. It is important to note that the rapid mobilization of resources in solid-state research after 1948 occurred in the university as well as in private industry. Thus, transistor technology was not building upon a vast *earlier* research commitment. That enterprise had, in fact, been extremely modest. It was the initial breakthrough of the transistor that gave rise to a *subsequent* large-scale commitment of scientific resources. J. A. Morton, who headed the fundamental development group that was formed at Bell Labs after the invention of the transistor, reported that it was impossible to hire people with a knowledge of solid-state physics in the late 1940s "because solid state physics was not in the curriculum of the universities."[24] As a result, Morton persuaded Shockley to run an in-hours course for Bell Labs personnel called "Solid State Physics of Semiconductors." Shockley's famous book, *Electrons and Holes in Semiconductors,* was a compilation of the materials used in that course. Shockley even ran a six-day course at Bell Labs in June 1952 for professors from some thirty universities as part of an attempt to encourage the establishment of courses in transistor physics. Clearly, the main flow of scientific knowledge during this period was from industry to the university.[25]

Thus, even when some basic research does precede a technological breakthrough, it is the establishment of a tangible link between technology and the specific field of science that is responsible for the great intensification of research in that field. A similar story concerns nuclear physics after the achievement of fission in 1938 and the awesome developments in military technology during the Second World War. At any point in time, the allocation of scientific resources is likely to be dominated by prior evidence of technological payoffs. This likelihood has been greatly increased in the twentieth century by the escalation in research costs and the consequent need to establish mechanisms for

[24] Morton, unpublished interview, p. 11.

[25] Even many years later, in centers of semiconductor activity such as Silicon Valley, it has been far from unusual for university courses in the solid state to be taught by "part-time professors from local industry." Ernest Braun and Stuart MacDonald, *Revolution in Miniature* (Cambridge University Press, Cambridge, 1978), pp. 126–7.

financing such research in the private and public sectors. But even in the late nineteenth century, before the great increase in research costs, the direction of scientific research was strongly influenced by technological achievements that promised high financial returns. The burgeoning of organic chemical research in the last third of the nineteenth century was largely a consequence of Perkin's successful synthesis of mauvine, the first synthetic aniline dye. The rapid expansion of basic research on the behavior of large molecules was a consequence of "the development by Leo Baekland, in 1909, of phenol-formaldehyde compositions which can be molded into any shape and hardened through molecular cross-linking by heating under pressure."[26]

VI

Why is the technological breakthrough so important to the direction of scientific research? Mainly for the obvious, but compelling suggestion of high potential payoffs, financial or social, to such research. It is important to realize that a major technological breakthrough really signals the *beginning* of a series of new developments of great importance, not their culmination. Here again, we are badly served by the stereotypical view of the temporal priority of basic research, of such research leading to or *culminating* in a technological breakthrough. In the most meaningful sense, the development of the transistor or the explosion of the first nuclear device or the first achievement of heavier-than-air flight is really the announcement of a new set of possibilities far more than their attainment. Indeed, it is tempting to *define* a major innovation as precisely one that provides an entirely new framework for technological improvements. That framework will often shape subsequent research for decades – just as we are now more than thirty years into the transistor revolution, and only the most bankrupt imaginations would suggest that we are close to exhausting its technological possibilities. Commercial success within that new framework requires numerous complementary inventions and the development of ancillary technologies, and these requirements also provide numerous focal points for scientific research. Thus, a great deal of scientific research is undertaken with the conscious intention of providing increments to knowledge that are perceived as being essential to the exploitation of the new technology.[27]

[26] Kranzberg and Smith, "Materials in History and Science," p. 25.

[27] As Harvey Brooks has noted in a discussion of the field of physics in the years following the Second World War: "The basic science was motivated by the necessity to generate ancillary technology to feed the development and exploitation of an initial invention, rather than vice versa." *Science*, 26 April 1968, p. 399.

Another basic reason why advances in scientific knowledge commonly occur after substantial technological improvements has to do with changes in the structure of economic incentives. A high-priced material is likely, *ceteris paribus*, to have a small number of industrial applications and will therefore generate only a limited interest – at least in the absence of some compelling performance advantage. On the other hand, as a material becomes cheaper as a result of technological improvements, it is more widely used, and its lower price causes it to be considered more seriously for further potential new uses. Thus, the major innovations in steel production in the 1850s and after created powerful incentives to perform extensive research on it. The cheapness of post-Bessemer steel made it possible to employ the metal for a wide range of structural purposes that were not previously economically feasible. But then it became extremely important to understand such things as the notorious variability of the metal, as well as its precise performance characteristics when subjected to a wide range of new stresses, tensions, and pressures in entirely new applications. Thus, the feasibility of employing steel on a large scale, *after* the innovations that led to its lower price and greater supply, led to considerable research and testing by new classes of users, as well as the steelmakers themselves. The possibility of making rails as well as other railroad equipment out of steel led to the creation of one of the earliest scientific laboratories in American industry – the Pennsylvania Railroad's laboratory, established in 1879 in Altoona. Here again, there are critical lines of causation running from the economic and technological realms back to the conduct of scientific research.[28]

Finally, and closely related, technological breakthroughs serve to validate the possibility of certain classes of phenomena and therefore also to heighten the likelihood of performing scientifically valuable, as well as technologically valuable, research at a particular site on the map of science. It is appropriate here to recall the great Rutherford's categorical denial of the possibility of releasing energy from the nucleus of the atom in the early 1930s, as well as Lord Rayleigh's denial of the possibility of heavier-than-air flight six or seven years before the Wright brothers took to the air – if only briefly – at Kitty Hawk. Technological achievements powerfully dramatize the scientifically interesting consequences as well as the purely financial benefits that may flow from research in specific areas, often by providing an empirical demonstration of the falseness of the conventional scientific wisdom.

[28] For details on the work performed at the Pennsylvania Railroad's laboratory at Altoona, see *The Life and Life Work of Charles Benjamin Dudley, Ph.D.* (American Society for Testing Materials, Philadelphia, 1911).

VII

There is another fundamental way in which technology shapes the scientific enterprise that I can only mention because it is, by itself, an extremely big subject. I refer to the development of techniques of observation, testing, and measurement – in short, instrumentation. Improvements in instrumentation, through their differential effects upon the possibilities of observation and measurement in specific subfields of science, have long been a major determinant of scientific progress. The full documentation of this assertion would be tantamount to a detailed discussion of the history of science over the past 400 years, and that subject is far beyond my competence. Let me make just a few points.

It would be easy to show, drawing upon the long history of the microscope (and, later, the electron microscope), the telescope (and, later, the radio telescope), and the recent histories of x-ray crystallography, the ultracentrifuge, the cyclotron, the various spectroscopies, chromatography, and the computer, how instrumentation possibilities have selectively distributed opportunities in ways that pervasively affected rates of scientific progress. However, to leave the discussion at that level would constitute a rather crude sort of technological determinism. I would suggest, first, that the relations between technology and science are much more interactive (and dialectical) than such a determinism would imply. For the decision to push hard in the improvement of one specific class of instruments will often reflect, inter alia, a determination to advance a particular field of science as well as an expectation that the relevant instrumentation is ripe for improvement. Furthermore, instrumentation technologies differ enormously in the range of their *scientific* impact. The linear accelerator and the ultracentrifuge are each relevant to a narrower portion of the scientific spectrum than, say, the computer. The computer, by contrast, is a general-purpose research instrument. It has had a pervasive impact upon a large number of disciplines, including the social as well as the physical sciences. Thus, different instruments may differ enormously in the specificity of their impact upon fields of science. Therefore, any attempt to establish tight links between progress in specific subfields of science and an associated field of instrumentation is doomed to failure.

Finally, it cannot be overstressed that improvements in observational capabilities were, by themselves, of limited significance until concepts were developed and hypotheses formulated that imparted potential meaning to the observations. The microscrope, after all, had existed for over 200 years, and many generations of curious observers had looked at strange little bacteria before Pasteur finally formulated a germ theory of disease in the mid-nineteenth century. After that, the microscope became far more significant for scientific progress than it had ever been before.

VIII

In conclusion, let me offer a restatement of the themes I have been trying to develop. Powerful economic impulses are shaping, directing, and constraining the scientific enterprise. These impulses are rooted in two facts: First, scientific research is a costly activity; second, it can be directed in ways that may yield large economic rewards. Industrial societies have created a vast technological realm that is very closely shaped by economic needs and incentives. This technological realm, in turn, provides numerous ways in which daily economic life has become closely linked with science. That realm defines the directions that promise large financial rewards and provides many problems and empirical observations that stimulate creative scientific research. These statements are supported by the increasing institutionalization of research in private industrial laboratories. It is fair to assume that decisions on the pursuit of science are subjected, in these profit-making firms, to a calculus of private costs and benefits.

If these things are so, economists may be guilty of excessive humility in treating science as an exogenous force that is not amenable to economic analysis. The factors that I have discussed have increased the ways in which, and the extent to which, scientific progress is being shaped by technological, and therefore economic, considerations. I believe that the industrialization process inevitably transforms science into a more and more endogenous activity by increasing its dependence upon technology. Technological considerations, I have argued, are a major determinant of the allocation of scientific resources. Thus, I suggest that a promising model for understanding scientific advances is one that combines the "logic" of scientific progress with a consideration of costs and rewards that flow from daily life and are linked to science through technology.

PART III

Market determinants of technological innovation

8 Technical change in the commercial aircraft industry, 1925–1975

David C. Mowery and Nathan Rosenberg

Judged by almost any criterion of performance – growth in output, exports, productivity, or product innovation – the commercial aircraft industry must be considered a star performer in the American economy. American commercial aircraft dominate airline fleets the world over, and the air transportation industry, a primary beneficiary of technical progress in commercial aircraft, has compiled an unequaled record of productivity growth since 1929. Along with this impressive record, however, the aircraft industry presents important anomalies in structure and conduct. Fierce price competition coexists with very high levels of producer concentration and significant product differentiation. The industry also has relatively little vertical integration; contractual relationships predominate in the pursuit of extremely complex and highly uncertain goals in price and performance.

Government policies regarding the commercial aircraft and air transportation industries have been partly responsible for this record of innovation and productivity growth. Government policy has influenced innovation in the aircraft industry through its impact upon the demand for aircraft, both military and civilian, and through direct support of research. The combination of high producer concentration and fierce price and quality competition also reflects the influence of government policy through the provision of both a market and research funding for military aircraft. This government role has also encouraged the development of a vertically disintegrated industry structure and an important role for subcontractors.

This paper provides a discussion and an assessment of government policy regarding the aircraft and air transportation industries. It also

This paper was originally published in *Technological Forecasting and Social Change*, 20, 1981, pp. 347–58. It is a much condensed version of "Government Policy and Innovation in the Commercial Aircraft Industry, 1925–75," written with David C. Mowery, to appear in Richard R. Nelson (ed.), *Government and Technical Change: A Cross-Industry Analysis* (Oxford: Pergamon Press, in press).

examines the impact of this policy upon the structure and performance of the aircraft industry. Of particular interest is the extent to which this apparently successful policy should be taken as a model for technology policy in other industries. Central to our argument is the assertion that government policy in this industry was successful because it provided substantial incentives for generating and adopting new technologies in commercial aircraft. The discussion begins with an assessment of the industry's innovative performance and some brief remarks on the general character of technological changes in aircraft. The development of the industry's structure and aspects of its current structure and performance are next considered. Finally, the structure and impact of government policy are discussed, with a concluding evaluation of the validity of this policy for application in other industries.

The performance and pattern of innovation

The most authoritative estimates of productivity performance in air transportation, those of Kendrick (1961, 1973), show it to have grown at an annual rate of 8 to 9 percent during the 1929–66 period, substantially higher than for any other American industry sampled. Various measures of aircraft product performance also display impressive gains for the 1925–75 period. Direct operating costs per seat mile dropped by over 90 percent between the era of the Ford and Fokker trimotors in the 1920s and the wide-body transports, utilizing high-bypass-ratio turbofan engines, in the 1970s. Passenger capacity and aircraft speed have increased by a factor of twenty during this period. Other performance measures, as well as such indices as passenger safety and comfort, also improved dramatically over this period.

Central to an understanding of the innovation process in the commercial aircraft industry is the great systemic complexity of the final product. The finished commercial aircraft is composed of a wide range of components for propulsion, navigation, and so on that are individually extremely complex. The interaction of these complex systems is crucial to the performance of an aircraft design, yet extremely difficult to predict from design and engineering data, even with presently available computer-aided design techniques. Uncertainty about aircraft performance is also exacerbated by the still modest state of scientific theory concerning the behavior of such key components as materials. Performance, in many cases, cannot be predicted definitively before the initial flight. This pervasive technological uncertainty has been and remains an important influence upon producer structure and conduct in the industry. Such uncertainty also introduces an additional dimension to the innovation process, learning by using.[1] A learning process occurs in the course of operating a new aircraft; as a result, operating costs are reduced. This process is not unique to aircraft but appears to be characteristic of complex products with elaborately differentiated, interdependent components. Operating cost reductions thus depend heavily upon learning more, during the operation of a new aircraft, about the performance characteristics of the system and its components, and therefore understanding more clearly its full potential.

This high degree of systemic complexity has facilitated innovation in at least one important way. The commercial aircraft industry has benefited greatly as a technological borrower. In addition to the often-remarked "spillover" effect of military designs upon civilian aircraft, aircraft companies have benefited to an unusual extent from technological developments in numerous other industries. Noteworthy examples are the metallurgical and materials industries, which have provided a wide range of new alloys and composite materials; the chemicals and petroleum industries, where important developments in fuels were achieved before World War II; and the electronics industry, which since 1940 has provided a stream of crucial innovations ranging from radar to airline reservation and navigational computers. The aircraft industry has benefited disproportionately from the interindustry flow of innovations that typifies the modern economy.

A final significant aspect of the aircraft industry is the need to achieve large production runs for a given aircraft in order to take advantage of learning curves and to defray high development expenses. Economies of scale and learning curves (the latter was first observed in the production of airframes) have a major effect on production costs and overall profitability. Such high development costs – which have assumed increased importance with the advent of the jet engine, reflecting the growing systemic complexity of aircraft technology – make it very important to maximize the production of a given aircraft design. This necessity, in turn, underscores the importance of the "family concept," in which a given aircraft, such as the Boeing 727, spawns a succession of modified designs, notably through stretching of the fuselage. Modern aircraft thus are designed so as to develop and exploit technological trajectories.

The development of industry structure

The evolution of the commercial aircraft industry may be divided into four unequal periods. In the 1920–34 period, military and commercial

[1] For a more complete exposition, see Rosenberg (1980).

aircraft production were gradually distinguished from one another, and peacetime military procurement came to play a role in airframe and especially engine development. In the immediate aftermath of World War I, the market for aircraft had collapsed with the cessation of military demand and a surfeit of war surplus aircraft. Production declined from 14,000 units in 1918 to 263 in 1922 (Holley, 1964). It slowly revived, particularly after the military announced plans in 1926 to maintain a total aircraft fleet of 2,600 by 1931, and the Kelly Air Mail Act of 1925 transferred the transport of airmail from the Post Office to private contractors. Also important during the 1920s was the increasing level and quality of research being carried out by the National Advisory Committee on Aeronautics, established in 1915. A series of mergers in the late 1920s created, for the first and only time in the history of the industry, several vertically integrated firms combining air transport, airframe manufacture, and engine production.

The onset of the Depression placed all manufacturers under considerable stress, but the airmail scandals of 1933 and the Air Mail Act of 1934 were decisive in dissolving these consolidated firms. Under the terms of the 1934 act, air transportation and aircraft manufacture had to be separated. The act also abandoned the goals of previous airmail legislation, which had been intended to promote the development of air transportation via subsidy, by specifying that minimum cost was to be the sole criterion for award of mail contracts.

From 1934 to 1940, four airframe producers and two engine manufacturers comprised the bulk of the civilian aircraft industry. During the 1930s, passenger rather than mail transport became the central activity of commercial air carriers. Throughout the 1930s, military production remained of great importance to the major commercial manufacturers. The military market was especially important to firms such as Boeing, Curtiss, Lockheed, and Martin, largely excluded from the commercial market after 1934 by the dominance of the Douglas DC-3. Although the military market was smaller (prior to 1938), the greater unit value of military aircraft enabled producers to avoid financial disaster.

During World War II, there was no real commercial aircraft industry. The heavy demand for aircraft spurred the development of a number of firms, previously minor participants in the commercial market, into potentially viable competitors. Proprietary control of military aircraft designs was reduced during this period, as cross-licensing of designs for maximum production was commonplace. In the rush to increase production, subcontracting came to play a crucial role. The large size of production runs also concentrated much greater attention on production engineering and the maximum exploitation of scale economies and learning curves. The in-house research and engineering capabilities of the major producers were greatly expanded.

During the postwar period, jet engine technology came to dominate commercial aircraft, causing substantial shifts in importance among airframe and aircraft engine manufacturers. Boeing and Douglas became dominant during the late 1950s and 1960s, whereas Lockheed, Martin, and Convair shrank into insignificance in the commercial market. An important consequence of the adoption of jet engine and electronics technologies was the spectacular rise in development costs for new commercial aircraft designs. From roughly $150,000 for the DC-2, development costs rose to over $100 million for the DC-8 and will exceed $1 billion for the Boeing 767. The rapid growth of these costs in effect means that an increasing proportion of the costs of introducing a new aircraft are incurred during the phase of greatest uncertainty concerning market prospects and technical feasibility. This increase in development costs and risks was partly responsible for the growing role of subcontracting within the industry.

Pricing, entry, and route structures in the commercial air transport industry were regulated by the Civil Aeronautics Board (CAB) during the entire postwar period. Price competition in transport was largely absent, but the presence of multiple carriers in important routes gave rise to intense service quality competition. Price competition among the airframe producers, however, remained strong despite increased producer concentration; the failure of the Convair 880 in the early 1960s, and the subsequent problems with the Douglas DC-9, were due in part to aggressive efforts by their producers to underprice the competition. More recently, the introduction of wide-body transports was marked by fierce competition between Douglas and Lockheed in both prices and delivery dates. The intense competition among airframe producers during the postwar period was responsible for several near-failures of major firms. A shotgun merger of McDonnell and Douglas Aircraft (aided by federal loan guarantees and Justice Department inaction) rescued Douglas from bankruptcy in 1967, and the collapse of Lockheed was averted in 1971 only by a federal loan guarantee of $250 million. Thus, to an unprecedented extent, the federal government was directly involved in determining the structure of the commercial aircraft industry in the 1960s and 1970s.

Industry behavior

The coexistence of high levels of producer concentration and fierce price competition makes the aircraft industry unusual in manufacturing.

One reason for this stems from the structure of the air transportation industry, a legacy of CAB regulation. The relationship between aircraft producers and airlines through 1978 was very close to bilateral oligopoly. The market for commercial aircraft was dominated by large orders from a small number (about four) of major trunk carriers. Airlines tended to have the upper hand in purchase negotiations, playing competing suppliers off against one another.

During the postwar period, the willingness of producers to undertake the expensive and risky development of new aircraft designs for which an insufficient market may exist reflects the importance of early delivery of new designs to airlines under the CAB regime. The advantages of having multiple suppliers of new aircraft designs also led airline customers to encourage competition in aircraft production. In addition, the federal government's unwillingness to allow a major airframe producer to go under undoubtedly encouraged producers to take sizable risks.

The aircraft industry, as presently structured, relies heavily upon contracting in the design, production, and procurement of complex capital goods; there is very little vertical integration. Production of new aircraft requires extensive negotiations between the airframe and engine producers, involving performance specifications and guarantees that are crucial to the success of a design, yet may be highly unrealistic at the time a contract is signed. One result is that major airframe producers have acquired great in-house expertise in engine performance, engineering, and evaluation, duplicating that of the engine manufacturers. But the subcontracting of new design production has also become very important in recent years, due to mushrooming development costs and the increasing complexity of aircraft components. Subcontracting has increasingly come to mean a sharing of risks. Rae (1968, p. 83) states that in the 1930s, subcontracting "constituted less than 10 percent of the industry's operations," but by the mid-1950s, 30 to 40 percent of the assembly work for the Lockheed Electra was subcontracted. Some 70 percent of the assembly for the Boeing 747 was subcontracted; in general, the same is true for the upcoming 767, as a means of spreading the commercial risk.[2] In addition, subcontracting in both military and commercial aircraft production has grown with the increasing complexity of aircraft components such as avionics. Major airframe producers do not have the requisite in-house competence to develop and produce many of these complex systems.

[2] According to *Aviation Week and Space Technology*, "The 767 subcontracting also will be devoted to sharing the risk. It will resemble the 747 situation in many respects, although in this case the major subcontractors will be required to assume a larger share of the risk, for potentially greater profits." (24 July 1978).

As development costs have escalated, large initial orders for a new aircraft design have become increasingly important. Producers must have a guarantee that at least a substantial portion of these costs will be recouped before the prototype is developed. The airlines placing these initial orders thus have great power to dictate the performance characteristics of a given aircraft. Since the route structures of the carriers vary greatly, the performance characteristics desired by each airline often differ substantially. The airline placing a large initial order may thus be in a position to influence the characteristics of a new generation of aircraft. As a result, the financial health and route structures of the airlines exert a major influence on the direction of technical change.

Large transactions costs exist within this industry structure. Extensive engineering staffs are maintained by airframe manufacturers, engine producers, and airline purchasers. Certain segments of the commercial aircraft market, notably short-haul aircraft, do not appear to have been well served by the innovations occurring during 1950–78.[3] The costs of negotiation and, not infrequently, litigation are high. Finally, the incentives for misrepresentation of performance characteristics and competition in price and delivery dates may reduce product safety.[4]

The reliance upon contractual relationships in many of the transactions involved in commercial aircraft production and procurement may occasionally result in severe impediments to the free flow of information and/or full revelation of details of design and performance (see Williamson, 1975, and Goldberg, 1976, 1977 for further discussion).

[3] In the early 1960s, the Federal Aviation Administration (FAA) attempted to develop a short-haul passenger transport capable of replacing the DC-3, then heavily utilized by local-service airlines despite its advanced years, lack of cabin pressurization, and low speed. A study of *Policy Planning for Aeronautical Research and Development*, prepared by the Library of Congress's Legislative Reference Service for the Senate Committee on Aeronautical and Space Sciences, noted that the FAA deemed action necessary because "While U.S. manufacturers had made a variety of studies, no design had been forthcoming . . . The key to starting the program appeared to be the need for a single order of at least 100 aircraft with the probability of at least 100 more. The local service airlines could not produce this order and only the DOD [Department of Defense] in Government could think in such quantities" (1966, p. 238).

[4] Eddy, Potter, and Page (1976) have argued that the Paris crash of the DC-10 in 1974 was due, in part, to such problems of incentives and information transmission, a proposition that received additional support from the pylon design and engine maintenance difficulties associated with the 1979 Chicago crash of the DC-10. In both cases, important information on changes in aircraft design and maintenance procedures was not transmitted effectively among the various independent organizations (McDonnell Douglas, the FAA, and the airlines) involved in operations, regulation, and maintenance. Incomplete revelation of possible product defects and/or design flaws may result from the conflicting interests of producers and consumers of complex, high-technology products, such as aircraft or nuclear reactors (as at Three Mile Island).

Offsetting these potential costs of market-mediated fabrication and procurement processes, of course, are the substantial benefits of competition among airframe and engine producers. It is unlikely that greater vertical integration in the commercial and transportation sector would have produced as rapid a pace of innovation, service quality improvement, and productivity growth.

Government policy and the supply of technology: the role of the NACA

The commercial aircraft industry is unique among manufacturing industries in that a government research organization, the National Advisory Committee on Aeronautics (NACA), has long existed to serve the needs of aircraft design. Similar research facilities, supported by both government and industry to carry out research on generic technological innovation, have recently been advocated for other industries by policymakers. The most frequent argument is that individual firms within a given industry have insufficient incentives and real disincentives (the free rider problem) in carrying out the basic research necessary to support innovation. The NACA is widely viewed by industry and government observers as a success in this regard.

The prewar performance of the NACA was achieved at a remarkably low cost, even by the standards of the time. Total appropriations for the NACA between 1915 and 1940 approximated $25 million. It is crucial to note, moreover, that the NACA carried out very little basic research during this period (see Constant, 1980). Prior to 1940, it functioned primarily to provide research infrastructure, and it made available extensive experimental design data and testing facilities, such as wind tunnels. This was a very important contribution, given the modest research resources of the industry prior to 1940, but it does not resemble the support of basic research frequently envisioned by advocates of government–industry research cooperatives. Following World War II, the NACA declined in importance as military funding of research mushroomed and airframe firms expanded their own research facilities. The NACA functioned as a sponsor of basic research in universities, and it was eventually absorbed into the larger National Aeronautics and Space Administration (NASA) in 1958.

Government policy and the supply of technology: military-sponsored research

The development of the first jet engine in the United States, drawing upon British assistance during World War II, was financed entirely by the military, reflecting both the perceived military urgency of the project and the lack of interest in developing such an engine by commercial firms prior to 1940. The development of commercial aircraft has also benefited substantially from military airframe development and production. With the advent of jet aircraft, airframe makers were able to apply knowledge gained in military projects to commercial design, tooling, and production. Development and tooling costs were reduced substantially. In certain cases, the costs of tooling for production of a commercial airframe were partially borne by government procurement contracts, as in the case of the 707 and KC-135.

Total research and development (R&D) spending in the aircraft industry rose by nearly 700 percent from 1945 to 1969; the most rapid growth was in the 1950–4 period, reflecting the large military funding during the Korean War. Expenditures rose from nearly $600 million in 1950 to more than $2 billion in 1954; 78 percent of this increase was accounted for by increases in military-supported R&D. Throughout this period, even in the late 1960s, the defense portion of total R&D expenditures never fell below 65 percent.[5]

The aircraft industry's contribution to R&D remained strikingly small in the late 1960s despite its rapid growth. It never accounted for more than 25 percent of total R&D spending and was below 20 percent for most of the 1945–69 period. However, industry expenditures accounted for an increasing share of nonmilitary research expenditures during this period, reflecting the growth of in-house research establishments and soaring development costs. From 42 percent of nondefense R&D spending in 1946, the industry share rose to nearly 64 percent by 1969.

The small portion of total industry research (both privately and publicly funded) that went to basic research is striking. The basic research share of total R&D expenditure was below 10 percent throughout the 1945–69 period, and the industry nonreimbursed share of this small fraction was below 10 percent. Public sources supported most of the basic research in the aircraft industry. The nonreimbursed industry share of applied research expenditures in 1969 was 34 percent.

Government policy and the demand for innovation

Government intervention and support to enhance the supply side of potential innovations in the aircraft industry has thus been substantial.

[5] Data in this and the next two paragraphs are derived from Booz, Allen and Hamilton Applied Research, Inc. (1971).

However, government policies have also been important in affecting the demand for innovation by the commercial aircraft industry. Consciously or not, the policies of the Post Office in the 1929–34 period and those of the CAB during 1938–78 influenced the structure and conduct of the air transportation industry, providing substantial incentives for rapid adoption of innovations.

Airmail transport was transferred from the Post Office to private contractors in 1925 following passage of the Kelly Air Mail Act. Bids were opened to private contractors on various mail routes; successful bidders were to be paid on a weight basis. During the ensuing five years, airmail postal rates were reduced by Congress, greatly increasing the volume of airmail, whereas payments to operators remained at their previous levels. The result was an increase in contractor profits. During this period, aircraft producers responded to the primary market, and aircraft such as the Boeing 40 were designed primarily for mail, rather than passenger, transport.

The McNary–Watres Act of 1930 and its administration by Postmaster General Walter F. Brown during the Hoover administration had the effect of developing a smaller number of large trunk carriers, who derived a far greater proportion of their revenues from passenger transport. The act changed the method of computing payments for mail carriage from a pound-mile basis to a space-mile basis, that is, payment was made whether or not mail was carried. In addition, extra payments were made to carriers utilizing multiengine aircraft, radio, and other navigational aids. The final major section of the act, which was to be its undoing, conferred substantial discretionary powers upon the postmaster general to alter or merge carriers when "in his judgment the public interest will be promoted thereby." Brown exploited his power to restructure air carriers to the fullest, working to develop a small number of financially strong transcontinental carriers that would constitute a sizable market for larger, more comfortable passenger transports.[6] His tactics produced

[6] The report of the U.S. Senate Judiciary Committee's Subcommittee on Administrative Practice and Procedure on *Civil Aeronautics Board Practices and Procedures* (1975) noted that "In May 1929 when Brown called together the airline industry to organize and rationalize it, 24 mail contracts were distributed among approximately 19 independent companies. Three companies received the dominant shares of the $11.2 million mail payments due at least to some extent to the excess payments required by the 1928 poundage method of contract payments: Boeing Air Transport (predecessor of United), National Air Transport (predecessor of United), and Western Air Express (predecessor of TWA).

"During Brown's tenure mail payments increased to $19.4 million and the miles of air mail routes increased from 14,405 to 27,678. By the end of 1933 all but 2 of the 20 air mail contracts were held by three large holding groups: United Aircraft & Transport Co.

a furor that resulted in the Air Mail Act of 1934, mandating divestiture by aircraft producers of subsidiary transport firms and placing the award of mail contracts on a per-ounce basis, to be given to the lowest bidder. Although the McNary–Watres Act was an inefficient mechanism, and Brown's administration of the act led to its demise, these policies coincided with rapid growth in air passenger traffic and the introduction of the monocoque fuselage air transports, the B-247 and DC-2, which were of great importance in the development of the commercial aircraft and air transportation industries.

Continued congressional dissatisfaction with passenger safety and regulatory policy in general led to the establishment of the CAB in 1938. By issuing operating certificates and overseeing fares, the board effectively controlled the pricing, entry, and route structures of the commercial airlines. These powers were used throughout the postwar period to prevent both entry into scheduled trunkline air transportation and price competition. As was mentioned under "The Development of Industry Structure," this regulatory environment created fierce competition in service quality. One result was a very rapid rate of adoption of new aircraft designs by the major carriers, based upon their belief that rapid introduction of state-of-the-art aircraft was an effective marketing strategy when price competition was not possible. The drive to be first with a new design strongly motivated major airlines to make early purchase commitments to airframe manufacturers as a means of achieving the earliest possible delivery. Service quality competition fostered rapid diffusion and adoption of innovations drawing upon government-supported research, and supported fierce competition among manufacturers. Repeatedly, purchases of new aircraft designs by one carrier were matched by a competitor, resulting in recurrent binges of new-equipment purchasing that left airlines burdened with heavy debts and excess carrying capacity.

Conclusion

In concluding this discussion of federal policy and innovation in commercial aircraft, we summarize our assessment of the role of the federal government in affecting innovation within the industry, and consider whether other industries could benefit from similar government policies. Although the innovative performance of the aircraft industry suggests

(United), Aviation Corporation (American), North American Aviation/General Motors (TWA and Eastern). These three flew 26,675 of the 29,212 air route miles and collected $18.2 million of the $19.4 million paid to air mail contractors" (pp. 202–3).

that this policy framework has been successful in some important respects, it is likely to be limited in its applicability to other industries. In view of certain other failings of both these policies and the commercial aircraft producers, the transfer of this policy framework to other industries may not be desirable.

The crucial aspect of federal policy throughout this fifty-year period is the fact that it has exercised an impact upon both the supply of and demand for innovation. Military support of new aircraft development provided important technical skills, knowledge, and innovations that could be utilized by manufacturers in commercial aircraft. Substantial research support for both military and civilian applications was also channeled through the NACA. Government demand for new designs, pushing at the outer limits of available technologies, was no less crucial in bringing about the rapid embodiment of new technical knowledge or isolated breakthroughs in some subsystem in a new aircraft design. The assurance of a market for a successful military aircraft gave manufacturers a strong incentive to pursue and utilize rapidly the technical and scientific knowledge acquired at federal expense. This assurance of the demand for innovative technologies is very important in understanding how technical breakthroughs were embodied so quickly in new aircraft. On the other hand, the modest success of such programs as the NASA technology utilization program or federally funded demonstration projects aimed at increasing the supply and availability of commercially useful knowledge, reflects in part the uncertainties about demand faced by the potential utilizers of this knowledge. The NASA program was also hampered by the often limited applicability of its technologies for civilian use. In the military aircraft market, which generated considerable commercial spillovers, such demand uncertainty was minimal.

On the demand side, the commercial aircraft market was also affected by government policies. We argued above that the McNary–Watres Air Mail Act, and the subsequent regulatory policies of the CAB, created a strong demand by the airlines for new aircraft embodying military-spawned innovations. Although the number of commercially unsuccessful aircraft indicates that the market was not assured, the regulatory policies motivated aircraft manufacturers to embody quickly new technological developments in innovative aircraft designs and pushed the airlines to adopt new aircraft designs as rapidly as possible.[7] To a lesser extent than in the military market, the existence of

[7] The episodes of the supersonic transport (SST) and Concorde illustrate the usefulness of a diffuse, rather than sharply focused, role for government in affecting the demand for a new technology. Both the SST and Concorde programs are examples of a misapplication of military procurement techniques to commercial aircraft development. Although it is

a strong demand for state-of-the-art commercial aircraft aided the rapid adoption of new technologies.

Publicly supported research is usually justified by labeling knowledge and information as a public good, arguing that the social payoffs for basic research may greatly exceed the private returns. Government support therefore is considered best when applied to the most basic forms of research. However, in the case of the NACA, established to provide research for the aircraft industry, basic research was notably absent. Constant (1980) argues convincingly that American firms failed to develop the jet engine prior to World War II largely because of the lack of theoretical work in aerodynamics and aeronautics in the United States, as opposed to Germany or Great Britain. The NACA's role prior to 1940, according to Constant, was primarily to provide testing facilities and empirical data, rather than to support advanced theoretical work in aerodynamics. Nonetheless, American firms were well placed to utilize the theoretical work in aerodynamics and the jet engine, most of which had been developed abroad, after World War II. The result was the 707 and the DC-8, the first commercially successful jet transports. Constant attributes the postwar U.S. dominance in jet aircraft to the extremely large, highly developed domestic airlines system that had evolved since the 1930s. Government agencies and policies, such as McNary–Watres and the CAB, because of their impact upon the demand for commercial aircraft, may have been as important as federal research support in the development of the postwar aircraft industry.

In promoting innovation, the experience of the commercial aircraft industry underlines the importance of affecting both supply and demand for innovation and technical knowledge.[8] Although this is a very general conclusion, with obvious relevance to technology policies in other industries, it is not clear that the policies utilized in the commercial aircraft industry are appropriate or applicable to other industries. Certainly, the resource costs of these policies in the aircraft industry have been substantial. High profits and federal research support in the

eminently sensible for the ultimate purchaser to specify in detail the operating and design characteristics of a given aircraft, the attempt to have designs for commercial application developed by an intermediary places great demands upon effective public–private sector communication and responsiveness. Also, it is important for decision makers to appreciate the commercial limits of success and to avoid the dangers of preoccupation with technical characteristics alone. Both the SST and Concorde designs paid little heed to operating costs, which probably would not have occurred if private airlines had controlled the design and development processes.

[8] Nelson and Winter (1977) and Mowery and Rosenberg (1979) provide analyses of the innovation process that emphasize the importance of linking both "market-pull" and "technology-push" forces.

development and sale of military aircraft have comprised an important government subsidy in the development and manufacture of new commercial designs. Carroll (1972) argues that government contracts were much more stable in volume, and yielded substantially higher profits, than commercial sales in the 1950s and 1960s. According to Carroll, the profitability of military sales may have led to fierce competition in commercial aircraft production and sales, including excessive duplication of development and tooling costs, as well as product lines, in some segments of the commercial aircraft market. As a result of this implicit subsidy, resources may have been inefficiently allocated. Further, we have argued that the competition between McDonnell Douglas and Lockheed may have reduced product safety. Finally, of course, there are the welfare costs to consumers of CAB regulation of air transportation, another element of the policy framework that has supported this high rate of innovation in commercial aircraft.

An aircraft industry policy paradigm may be relevant in the area of technologies for reducing emissions of pollutants and carcinogens in automobiles and industry. Here, the performance characteristics of the technologies mandated by federal regulation could be clarified so as to make the demand for innovation clear and unambiguous. Coupled with increased government funding of research in this area, policies could be developed that would affect the supply of technical knowledge and innovations, as well as the demand for new emissions control processes, so as to improve the state of the art. Another area in which such an approach may be useful is energy technologies. Here, the government currently funds research extensively, in contrast to the area of emissions control technologies. However, it has done little to provide a clear and stable demand for energy technologies with specific cost and performance characteristics. (Indeed, until the recent moves to remove price controls on oil and natural gas, government policies discouraged the application of new technologies.) By making commitments to purchase certain forms of energy at a guaranteed price – for example, synthetic fuels for a strategic petroleum reserve – or the output of certain technologies with specific cost or performance characteristics – for example, solar energy sources meeting announced criteria – federal policies could provide a more effective set of "market pulls" in addition to the current "pushes" from extensive research funding. The essential requirement is to design policies that affect both supply and demand.

References

Booz, Allen, and Hamilton Applied Research, Inc. (1971). *A Historical Study of the Benefits Derived from Application of Technical Advances to Commercial Aircraft,* prepared for the joint Department of Transportation–NASA Civil Aviation R&D Policy Study. U.S. Government Printing Office, Washington, D.C.

Carroll, S. L. (1972). "Profits in the Airframe Industry." *Quarterly Journal of Economics*, November.

Constant, E. W. (1980). *The Origins of the Turbojet Revolution*. Johns Hopkins University Press, Baltimore.

Eddy, P., Potter, E., and Page, B. (1976). *Destination Disaster*. Quadrangle Press, New York.

Goldberg, V. P. (1976). "Regulation and Administered Contracts." *Bell Journal of Economics*, Autumn.

(1977). "Competitive Bidding and the Production of Precontract Information." *Bell Journal of Economics*, Spring.

Holley, I. B. (1964). *Buying Aircraft: Material Procurement for the Army Air Force*, Vol. 7 of the Special Studies of the U.S. Army in World War II. U.S. Government Printing Office, Washington, D.C.

Kendrick, J. W. (1961). *Productivity Trends in the United States*. Princeton University Press, for the National Bureau of Economic Research, Princeton.

(1973). *Postwar Productivity Trends in the United States, 1948–1969*. Columbia University Press, New York.

Mowery, D. C., and Rosenberg, N. (1979). "The Influence of Market Demand upon Innovation: A Critical Review of Some Recent Empirical Studies." *Research Policy*, April.

Nelson, R. R. and Winter, S. G. (1977). "In Search of a Useful Theory of Innovation." *Research Policy*, January.

Rae, J. B. (1968). *Climb to Greatness*. MIT Press, Cambridge, Mass.

Rosenberg, N. (1980). "Learning by Using." Unpublished manuscript.

U.S. Senate Committee on Aeronautical and Space Sciences. (1966). *Policy Planning for Aeronautical Research and Development* U.S. Government Printing Office, Washington, D.C.

U.S. Senate Judiciary Committee, Subcommittee on Administrative Practice and Procedure. (1975) *Civil Aeronautics Board Practices and Procedures*. U.S. Government Printing Office, Washington, D.C.

Williamson, O. W. (1975). *Markets and Hierarchies*. The Free Press, New York.

9 The economic implications of the VLSI revolution

Nathan Rosenberg and W. Edward Steinmueller

To examine the prospective economic impact of a development in electronics – VLSI (very large scale integration) – we must first focus explicitly upon the variables which mediate between the technological and the economic realms, that is, between technical feasibility and commercial success.

Cost versus performance

Within the purely technological realm one can deal with performance criteria, or with the solution of mechanical or other engineering problems, in a relatively unconstrained way. So long as cost considerations are not critical, one can conceive of a wide range of technically feasible alternatives for improving the speed of an airplane, increasing the strength of a bridge or the hardness of a material, or providing an electricity supply for a specific location. However, technical success (or purely mechanical performance measures) is only a necessary and not a sufficient condition in establishing social usefulness. The bankruptcy courts provide evidence of the perils of an excessive preoccupation with technical feasibility or performance, and it is notorious that the overwhelming bulk of patents are never commercialised.

For a technological improvement to exercise a significant social impact, it must ordinarily fulfill additional criteria. Specifically, it must combine design characteristics that will match closely with the needs and tastes of ultimate users, and it must accomplish these things subject to the basic economic constraint of minimising costs. The major excep-

This paper was originally published in *Futures*, October 1980, pp. 358–60. It is based on a paper presented to the session entitled "The Impact of Very Large Scale Integration" at the American Association for the Advancement of Science annual meeting, San Francisco, 4 January 1980.

Nathan Rosenberg gratefully acknowledges the financial support he received from the National Science Foundation during the writing of the paper.

tions, in our society, are defence and related aero-space expenditures, where enormous priority is placed upon performance and where cost considerations are regarded as secondary.[1] That these sectors should be continually plagued by 'cost overruns' is, accordingly, hardly surprising. The accepted trade-offs between cost and performance criteria are such that defence contractors systematically structure their development activities to give priority to improvement of performance. In addition, the penalties for exceeding cost estimates or failing to meet delivery dates are not nearly as severe as the penalties for failing to meet performance specifications.

By contrast, the long-term impact of technical improvements in the civilian sector may be expected to turn upon the attainment of low cost levels that will render the product superior to available substitutes on a cost basis, or that will provide a superior product at a cost that is at least not prohibitively expensive by comparison with lower-performance substitutes. Higher performance is often attainable at a higher price in civilian markets. However, to choose the optimal combination(s) of price and performance at which a firm should aim calls for considerable knowledge of market conditions as well as business acumen of a high order in making decisions on timing. At the extreme end of the spectrum, excessive preoccupation with the improvement of certain performance characteristics, without concern for cost, is a recipe for financial failure. The Concorde, in many respects a brilliant engineering achievement, is an obvious commercial disaster – as is immediately apparent from a comparison of its costs per passenger mile with those of wide-bodied jets.

Similarly, the history of nuclear power over the past 30 years is the story of a spectacular scientific and technological breakthrough but of only a relatively modest economic achievement. In the early euphoric days of nuclear power there were advocates who spoke confidently of a future ability to produce electricity so cheaply that it would no longer be worth installing meters to measure consumption. The nuclear industry has been plagued by numerous unanticipated safety and environmental problems in recent years, but its modest growth rates were, for long, due simply to its difficulty in competing on a cost basis with conventional fossil fuels.

[1] The provision of medical services should perhaps now be regarded as occupying an intermediate category between the military sector (where costs are secondary) and the conventional civilian sector (where they are the dominant consideration). The rising costs of medical services in recent years are attributable, in part, to an underlying change in social attitude: High-quality medical care is now seen as a universal right rather than merely a purchasable service.

Solar energy, despite having many attractive properties and articulate spokesmen, remains too expensive for most purposes even at today's high fossil-fuel prices, and is unlikely to be widely utilised unless improvements bring order-of-magnitude reductions of cost.

Automation, the electronic control of production processes, differs from the other innovations we have mentioned – for many purposes, its technical feasibility has yet to be established. When it does become technically feasible across a wider spectrum of industrial activities than is the case at present – and that point may now not be far in the future – it will, like all civilian innovations, still have to run the gauntlet of commercial considerations that we have been emphasising. In the civilian sector, the diffusion of an innovation is dictated by economic performance, that is, cost per unit of output.

As circuit-element size shrinks, capital costs increase

It is our contention that, while VLSI represents a dramatic step forward in chip complexity, it is fundamentally a step within the parameters of the existing IC (integrated-circuit) industry; this new technology is governed by a similar set of parameters and cost factors.

In the IC industry both the cost of devices and their performance have changed dramatically over the life cycle of individual products and with the advent of new product types. The primary factor responsible for these changes in cost and performance has been the process technology underlying IC production. The dominant trajectory of this technology's development has been the reduction of circuit-element size and hence increases in circuit-element density in a chip. A greater number of circuit elements becomes possible every time circuit elements and their interconnections can be decreased in size.

In this article, we define VLSI as IC devices having connection lines that are less than 1.5 μm in width, that is, devices with at least 100,000 circuit elements. While some ICs now in production meet this definition, the processes by which they are produced are not yet widely used.[2] Improvements and innovations in processing technique are expected to lead to greater commercialisation of VLSI ICs.[3] Thus the term VLSI technology in this article denotes both the continuing increase in numbers of circuit elements resulting from improvements in processing technology, and devices with more than 100,000 circuit elements created by using this improved technology.

[2] J. Lyman, "Lithography chases the incredible shrinking line", *Electronics*, April 12, 1979, pages 105–16.

[3] Ibid.

The pace of advance in eletronics process technology has been extraordinarily rapid in comparison with that of other technologies. Osborne observes that, if transport technology had progressed from stagecoach to Concorde as rapidly as electronics technology since the transistor, Concorde would be able to "carry half a million passengers at twenty million miles per hour."[4]

New IC products tend initially to be both more expensive and of higher performance. But, subsequently, the price of these higher-performance devices falls so that the all important cost per unit of performance dramatically declines as each new product reaches the mature stage of the product cycle. The successive improvements in cost per unit of performance have been as dramatic as the advances in process technology. Osborne concludes his stagecoach/transistor comparison by noting that: "a ticket for a Concorde flight would have to cost less than a penny if it were to compare with the rate at which microelectronics has gotten cheaper".[5]

Declines in cost per unit of performance of this magnitude reflect high investment and rapid innovation. Increasing IC component density has been made possible by more complex processing equipment. Process equipment has become much more expensive as a result of its increasing complexity. Thus, the capital or fixed costs of IC production have increased substantially. The cost of a wafer fabrication plant has increased from about $2 million in 1970 to $50 million in 1979.[6] At the same time, most IC prices have declined almost as fast as costs per unit for the new high-capacity plant. The consequence of these two trends is the necessity for IC companies to operate at very high volumes. More than ever, high volumes are required to amortise higher capital equipment costs. And if it is to generate products that are competitive with other high-volume producers, the firm must constantly upgrade its fabrication plant to produce the new denser ICs. The investment necessary to generate $1 of further sales has increased from $.30 in 1970 to $0.70 in 1979.[7] Historically, this problem has been compounded by rapid technical change that makes process equipment obsolete and hence further increases capital costs.

In summary, the process improvements leading to greater circuit-element density have led to dramatic increases in IC chip capabilities.

[4] A. Osborne, *Running Wild – the Next Industrial Revolution* (Berkeley, CA, Osborne/McGraw-Hill, 1979), page 162.

[5] Ibid., page 163.

[6] C. K. Mick, "Some current data and trends on semiconductors", processed, Decision Information Services Limited, Palo Alto, CA, 1979.

[7] Ibid.

But this trajectory of process improvement has substantially increased capital costs in an industry where price has fallen over time. The results of these trends are the increasingly pressing necessity to produce at large-scale volumes, and the higher costs of entry into the industry.

High-volume production favours large established firms

The 'requirement' of large-scale production to achieve unit-cost economy and overall profitability for further company growth has had two extremely important implications. First, it has affected the industrial organisation of the IC components industry. And second, it has been a major influence on the design of products.

The industrial organisation of the IC industry has changed since the 1960s when firms were founded by a few talented entrepreneurs whose investment of several thousand dollars provided the basis for millions in future sales. The primary reason for the demise of this sort of entrepreneurship has been the growth in capital requirements for the production capacity necessary to be a contemporary competitor. It is difficult to assess whether financing of entry has become more difficult. But it is possible to say that the source of financing appears to have changed. Venture-capital groups, which were important in the earlier growth of the industry, do not have the resources required to finance a prospective entrant in today's market. The consequence has been that the only entrants since 1974 have been 'captive' supply houses to 'sell' only to their owner or newly established electronic divisions of major nonelectronic corporations.

VLSI will probably accelerate the existing trends toward the increase in the size of firms in the industry. Certainly it is already apparent that VLSI production technology will make initial investment requirements much higher. However, the rapid continuing pace of technical progress in the industry makes it extremely risky to build large plants in order to exploit scale economies. This is so because, by the time the plant becomes available for volume production, it may prove already to be obsolete. Thus, a critical correlate of rapid technical change is an intensification of risks for potential investors. For a new entrant, who needs an even greater period of time to undergo the necessary learning experiences in mastering an unfamiliar technology, the risks are far more formidable and the prospects of success far more doubtful.

Standardisation and the drive for economies of scale

The necessity for large-scale production has had one very important implication for product choice: standardisation. IC firms have been compelled to standardise designs and to agree on industry standards of connection and communication. Standardisation of devices often occurs through several companies following a single set of specifications. These specifications may be promulgated by an important user such as the military, or more commonly by a company that takes the initiative. In either case, standardisation assures system designers that a device will be available because several companies will be producing it and it provides them with specifications for the device so that they can begin to design systems with it even before delivery.

Since systems require a number of IC components to be electrically connected on a printed-circuit board and to exchange electronic signals with one another, connection standards are necessary to provide a means of assuring successful linkages between components, even if produced by different manufacturers. Connection standards include conventions such as signal timing, error correction codes, and even more fundamental information encoding rules such as the ASCII (American Standard Code for Information Interchange) for alphanumeric information.

Standardisation and connection standards may seem purely technical details to the casual observer, but in fact they reflect the importance of achieving economies of scale. Components that failed to meet such standards could not economically be used in many systems regardless of their individual advantages. A firm with a nonstandard component would be forgoing the bulk of the market and would be sacrificing the attendant economies of scale. Most firms would be likely to abandon attempts to design products that could not meet such standards, opting instead to produce ICs that could achieve economies of scale. The necessity of large-scale production therefore implies that standardisation and connection standards will continue to be important in developing mass markets for IC components.

Cheaper, larger, faster memories

The economic requirement of large-scale production will also influence the type of device and arena of competition chosen for near-future VLSI ICs. As Noyce has noted, the major area of potentially large markets and attendant scale economies for VLSI technology is the IC memory market.[8] Increases in capacity and speed resulting from improvements in production technology have, in the past, resulted in a continuing

[8] R. N. Noyce, "Hardware prospects and limitations", in M. L. Dertouzos and J. Moses, eds., *The Computer Age – a Twenty-Year View* (Cambridge, MA, MIT Press, 1979), pages 321–37.

decline in the cost of memory per unit of information and speed. Since memory capacity, and to a lesser extent speed, are vital to an extremely broad class of applications, every effort will be made to continue this trend of increasing capacity and faster speed at ever lower cost. The extent to which VLSI becomes a dominant technology in the IC components industry will largely depend on the amount of price reduction achieved by the coupling of process improvement and scale economies, which are dependent on the size of the market.

Every new technology must compete with and displace past technologies by becoming more cost effective. In the case of VLSI, LSI memory technology will be a competitor until VLSI-related process innovations are successfully commercialised. That is, until VLSI process technologies reduce costs per unit of complexity below LSI costs, improvements in LSI process technology may limit the extent of commercial success of VLSI devices. The specific nature of this competitive process will shape the initial applications of VLSI technology. But, once developed, commercially successful techniques for creating extremely high circuit-element density for VLSI memory will be used to produce other VLSI circuits.

Flexible building blocks present new opportunities

The advances of the 1960s in circuit-element density permitted the invention of the microprocessor. Experience with the microprocessor in the 1970s may serve as some guide to the criteria for design of logic devices that the process advances of this decade have made possible. Initially, the microprocessor was regarded as a novelty of limited applicability – because, when compared with existing systems, it seemed but a small and inferior 'computer'. But designers soon realised that the microprocessor could be programmed to simulate any one of a vast number of logic circuits, and in fact was a generic circuit of extremely broad applicability. Intel and other manufacturers of the device found they had an enormous market and were able to set up large-scale and hence profitable production operations. The programmability of the microprocessor represented an opportunity to employ a single device as a building block in a vast number of different applications by devising different instruction sets.

Similarly, the more complex ICs made possible by VLSI offer opportunities to invent devices with even greater capabilities and flexibility for use as system building blocks. These devices, if successful, will generate the even larger volume of production required to amortise the higher capital equipment costs that VLSI will surely entail. Thus, while a single economic objective, the creation of cheaper memory with greater capacity, may create a new technique, VLSI, this technique may serve, in turn, to create an entirely new opportunity for innovation.

Process innovations have been extremely important in the electronics industry. They have led to technical and economic consequences unanticipated by the designers and of far greater importance than those that induced the innovation in the first place. Historically, there is considerable basis for concluding that this is how the fate of many a major innovation unfolds. In this way, the development of electronics is similar to that of other technologies.

The impact of innovations: a review of underestimates

We have suggested that the economic impact of an innovation such as VLSI must be assessed by 'translating' technological variables into meaningful economic counterparts. However, there is often a gross failure to anticipate many of the eventual specific end uses of an innovation. A great deal of creative intelligence and social imagination is necessary to perceive new possibilities at that vague interface between new technological capabilities and social needs.

It has been said that when Edison invented a technique for the reproduction of sound, he thought that its main application would be for recording the death-bed wishes of elderly gentlemen. When the radio was being developed, it was thought that its uses would be confined mainly to communicating with remote or inaccessible places where communication by wire was not feasible – for example, on mountain ranges or on ships at sea. The immense possibilities for commercial broadcasting and the vast potential of the innovation as a medium for entertainment as well as news broadcasting and information dissemination were, astonishing as it may seem in retrospect, almost totally unforeseen.

For a number of years after the emergence of the oil industry in the USA, following the successful drilling in Titusville, Pennsylvania, in 1859, oil was eagerly sought, not as an energy source at all, but rather as an illuminant. The main output of the oil industry in its early days was kerosene. It was only toward the close of the 19th century, with the development of the internal combustion engine, that the potential of oil as a new power source was capable of realisation.

Closer to the subject of this article and less than 30 years ago, in the early days of the computer, the serious prediction was made that the future computing needs of the USA might be readily accommodated by a total of a dozen large computers. There are, we exphasise, good reasons for this limited foresight. How a given invention will fit into a large and complex social system, the applications that will be devised for it, and the alterations that will be generated by its presence are all extraordinarily difficult to predict.

The impact will be highly sensitive to the income level of the society and related aspects of its social structure. An invention that may have a large acceptance and impact in a rich and urbanised society may have no impact at all in a poor agricultural society with low population densities. The commercial aircraft industry emerged in North America and Western Europe, where there were widely scattered urban centres with relatively rich populations. Even if the necessary technological capabilities had existed in Northern Africa, it is most unlikely that commercial aviation could have originated there.

In addition, inventions are heavily dependent upon the availability of essential complementary inputs, and the willingness and ability to supply those inputs. The North American household would be a very different place in the absence of electricity. The automobile would have exercised a far more modest impact upon US society in the absence of an elaborate network of paved roads provided by the government, and gasoline stations and repair shops provided by private industry.

The growth of the computer industry will plainly be hampered by the difficulty in developing complementary software to better exploit the remarkable potential of even the presently available hardware. Indeed, it is possible to conceive of a prolonged flow of new and improved products based merely on the development, for hardware already on the shelf, of new software.

The identification of human needs

Innovation entails a subtle combination of technical sophistication with the identification of specific but unsatisfied human needs. The phenomenal success of CB (citizen's band) radios at the consumer level and the minicomputer in industry and business relied on the extension and adaptation of technologies to new social contexts. New technologies may also be combined with older ones in ways that are extremely difficult to anticipate – for example, the application of the minicomputer to the problems of controlling industrial processes. Or there may be a failure to anticipate the immense increase in the impact of an innovation that can flow from drastic reductions in cost or improvements in performance. Again, the remarkably rapid growth of the minicomputer in business and industry in recent years underlines this point forcefully.

But, beyond this, major technological innovations such as the integrated circuit and its projected extension into VLSI really constitute structures upon which it is possible to build in many different directions. The steam engine in its original Newcomen form was 'merely' a device for pumping water out of mines. A succession of improvements in the device rendered it a feasible source of power for textile factories, iron mills, and an expanding variety of industrial establishments. In the course of the 19th century the steam engine was used to produce a new, generalised source of power – electricity – that in turn satisfied innumerable final uses to which steam power itself was not efficiently applicable. Finally, the steam turbine displaced the steam engine in the generation of electric power, and the special features of electricity – its ease of transmission over long distances, the capacity for making power available in 'fractionalised' units – and the far greater flexibility of electrically powered equipment spelled the eventual demise of the steam engine.

The economic impact of VLSI is unclear because, as in the example of the steam engine, it is difficult to anticipate the *direction* in which subsequent building upon this new structure will occur. New inventions are usually perceived as if they were merely superior versions of older, established technologies when, in truth, they are really the basis for entirely new technological developments. In electronics, for example, the transistor was thought of for some years as merely a greatly improved substitute for the vacuum tube (thermionic valve).

Who in the 1770s, when James Watt was establishing the widespread commercial feasibility of the steam engine, could have anticipated that a major field of its application would be in new forms of transport, or that it would one day be used to generate electricity in order to illuminate houses? (Electricity, in turn, was introduced to provide light but later became a new industrial power source and then the basis for household appliances.)

The later directions that the steam engine took were powerfully influenced by new inventions undreamed of in Watt's time – such as electricity – but also by the emergence of certain crucial needs in specific social and geographic contexts. Thus, American leadership in the development of the steamboat reflected, in part, the extremely high economic payoff to a growing nation of continental proportions, rich in agricultural resources and well endowed with an internal system of natural waterways. The visionaries of the early 19th century – Oliver Evans, John Fitch, Robert Fulton, James Ramsey and John Stevens – all correctly anticipated that, in the American context, there would be great economic benefits in the successful application of steam power to the navigation of western rivers. The speed and vigour with which Americans, at a later date, applied and exploited the railroad, after its invention in Britain, represented a similar response to the same structure of economic opportunities. On the other hand, even if the automobile had become technically feasible and reasonably cheap in 1800 rather than 1900, it is unlikely that conditions in 1800 would have permitted wide-scale adoption.

These historical allusions demonstrate what we regard as a fundamental point: The economic impact of VLSI, like that of other innovations of great potential, will depend upon our success in linking this new technological capability with the solution or the amelioration of some of our most pervasive prospective problems. It is clear that, at the very least, VLSI constitutes a technological building block that, if it meets even modest expectations, will allow the historical trend of increasing chip complexity, accompanied by reductions in cost per unit of complexity, to continue.

VLSI diffusion depends on knowledge as well as costs and capabilities

How are these VLSI building blocks likely to be applied to useful purposes? There are two important links between social needs and cost and capability advances in IC components. The first such link is the system producer or 'original equipment manufacturer' who utilises components to build systems for a variety of applications. The second is the application user for whom electronic systems and the components upon which they are based must serve some useful purpose. Larger markets for electronics technologies are built by enhancing existing links and developing new ones, by developing new products or adapting existing products to new uses.

We have already analysed some of the factors within the IC-components industry that will influence the cost and nature of future IC products, including VLSI. A more difficult issue is the economic process by which new IC techniques will diffuse to other industries and thereby generate significant productivity improvements elsewhere. This diffusion is perhaps most influenced by the cost and nature of products available from the components industry. But it also requires considerable knowledge about the problems of potential user organisations, the extent of commonality among user needs, and the appropriate amount and nature of software needed by the user. For these reasons, the diffusion of electronics technology is knowledge intensive. The creation and transfer of this knowledge both within and between organisations will be an important determinant of the extent to which certain innovations will be adopted and will exert influence on the underlying economic factors.

Adoption will be selective and incomplete

Rapid technological progress in the data-processing and communications industries forces each company to keep abreast of competitors by adopting improved technologies, that is, it produces rapid diffusion of new components technologies. But when the incremental advance in cost saving or performance in such industries is compared to that in other industries that have not yet gone electronic, the possible (but yet to be adopted) advance does not seem as great as might be expected.

In some industries, the improved electronics technology may not offer across-the-board economic advantages over the existing technology. For example, numerically controlled machine tools became available on a commercial basis in the mid 1950s. But the present state of the art offers economic advantages only for intermediate-size production lots, and not for large or small ones. Numerically controlled machine tools therefore took more than 15 years to capture 25 percent of machine-tool sales, whereas in 15 years semiconductors achieved no less than 85 percent of the combined semiconductor/vacuum tube market.[9]

The rate of diffusion of new technologies is intimately linked to the speed with which they come to offer distinct economic advantages over old technologies, which may continue to be improved, or to offer economic advantages for specific uses. The economic advantages of the new technology will therefore be selective and incomplete, rather than across the board. Even vacuum tubes have not been completely displaced by semiconductors.

The gatekeepers who evaluate new technology

Thus, economic factors exercise a pervasive influence over the rate of diffusion of new technologies. One of the most important determinants of innovation is that of evaluation by potential adopters – which depends on the ability to anticipate uses and requires intimate familiarity with the technology.

It should be apparent that people with such skills are not uniformly

[9] E. Mansfield, "The diffusion of eight major industrial innovations", in M. E. Terleckyj, *The State of Science and Research: Some New Indicators* (Boulder, CO, Westview Press, 1977), pages 85–121.

distributed through the labour force. Instead, they are concentrated in those areas where they have historically gained the largest return on their knowledge, the computer and communications industries. These technological 'gatekeepers' are crucial to the diffusion of new electronic technologies because of their unique role as intermediaries. They have developed an understanding of their organisations' needs and perhaps have ideas for technological solutions that are feasible. Components producers, on the other hand, have only a limited knowledge of their ultimate, end-use customers' needs. Systems producers have a more detailed knowledge of their users' requirements but for many applications, especially in industries where little use is currently made of electronic technology, the initial contact is likely to be made by the gatekeeper.

The gatekeeper is usually unable to present the system designer with a complete specification of the necessary tool for solving the organisation's problems. Hense, an increasingly important activity of systems houses and independent consultants is to advise and offer technological solutions to gatekeepers from other organisations. This form of technology transfer, though poorly understood, will greatly influence the rate and direction of diffusion of electronics technology.

Enhancement of the links that already exist between component producers, system designers, and final users requires more effective design of systems that can be mass marketed. The development of such systems is closely related to the knowledge that the gatekeeper attempts to bring to bear on the problems of the organisation. To be successful, system producers must translate their experience with customers into a set of criteria for designing new products to meet the needs of a very large number of users. They also must relate the requirements of the product to available technology. Thus system producers will also play an important intermediary role in relating problems to potential technological solutions.

Certain technical advances open up vast new areas of applications and generate spectacular increases in sales. This same opportunity is a problem in the relationship between systems and components producers. In an era of increasing circuit complexity, if systems producers are very successful at opening vast new areas, they encourage imitation by components producers who may be able to produce the electronic part as a 'system on a chip'. Since most system designs are not patentable, this raises serious questions about the adequacy of incentives to produce new systems. A closely related question in new system design is how software investment decisions are to be made in an era of declining hardware costs.

The choice of appropriate software level – a block on diffusion

Software development shares many of the problems of any R and D activity. There is the 'free rider' problem – it is difficult to protect the product from copiers once it has been produced. Economies of scale may not exist. Output is well nigh impossible to measure and hence productivity is difficult to determine. If the project fails there is probably no useful product at all. All of these factors contribute to the difficulty of making correct decisions about how much software to create.

Yet software is vital to modern electronic systems and determines many of the important features of the final system product. The difficulties of determining an appropriate software level for investment directly affect the functionality and usefulness of electronic products. No easy solution to this problem seems to exist and we believe it is a fundamental limitation to the diffusion of complex electronic technologies. For example, automation, the exercise of electronic control over production processes, requires a great deal of software investment. But problems in assessing the size of the software design task and its benefits have made it very difficult for gatekeepers to choose an appropriate level of software investment – thus slowing the progress of automation.

Three factors are important in the diffusion of new components:

- the supply and talents of technological gatekeepers,
- the capability of systems producers to bridge the gap between technological capabilities and a common denominator of user needs, and
- the increasingly urgent need for more software investment.

All three demonstrate the fact that the diffusion of a new technology is a highly knowledge-intensive undertaking.

Electronics, wisdom, incentives, and imagination

If VLSI meets the most optimistic expectations for cost and performance improvements of IC components, numerous opportunities will exist to tackle urgent contemporary concerns. Although the technological capability will exist, the imaginative application of this capability to areas of major social concern is by no means a certainty.

To what extent can this capability be applied to the obviously urgent concerns of the 1980s: the energy crisis in its many ramifications, our determination to achieve higher standards of pollution abatement, and the virtual stagnation of productivity growth? Each of these concerns takes us far beyond the realm of technology alone. Each is shaped by a complex structure of incentives, legal and regulatory as well as economic, that influence the behavior of individuals and firms.

Nevertheless, it seems clear that there are innumerable ways in which our growing electronic capabilities may be employed to deal with our larger social concerns. The extent of our eventual success in achieving such applications will turn upon incentives, wisdom, and imagination as well as upon purely technological virtuosity.

10 The influence of market demand upon innovation: a critical review of some recent empirical studies

David C. Mowery and Nathan Rosenberg

Introduction

Until quite recently, economists devoted little attention to the factors that influence the rate and direction of innovation. Much of the formal economic theory on technical change is really concerned with the description of the consequences of technological innovation at a very high level of aggregation or abstraction. (We have in mind here the works of such economists as Harrod, Hicks and Denison.) Little consideration has been paid to the study, at a less aggregated level, of the specific innovative outputs of industries and firms, and the forces explaining differences among industries, firms, and nations.[1] Serious empirical work on biases and inducements in the innovation process, at an industry or firm level of analysis, is even more conspicuously lacking. A recent theoretical literature has been rather inconclusive – with results turning to a distressing degree upon the nature of the assumptions and the particular form of characterization of rational behavior.[2]

An understanding of the innovation process at this level is, however, essential. Obviously, an understanding of the influences that motivate innovation, and channel its direction, is necessary if government intervention is to be successful in increasing the production of useful innovations in specific areas. Awareness of the need for empirical analyses along these lines has been joined to a sense of greater urgency concern-

This paper was originally published in *Research Policy*, 8, April 1979, pp. 103–53.

[1] One of the most important recent attempts along these lines was the historical work of Habakkuk (1962), which gave rise to an interesting discussion. (See also David, 1975, ch. 1.)

[2] See Samuelson (1965), Kennedy (1964), and Salter (1966) for a sample of this extensive literature.

ing the need for conscious intervention on the part of the government, in view of the growing concern with problems of slower growth in productivity and income, environmental and urban decay, and resource, energy and balance-of-payments constraints that appear to face many countries in the 1970s and beyond.

A number of new empirical studies of innovation have, accordingly, been funded by governmental bodies, while the conclusions of older ones have been exhumed and reinterpreted, in an effort to inform the policy-making process. These studies, and summaries of earlier ones, have led a great many scholars and officials to the conclusion that the governing influence upon the innovation process is that of market demand; innovations are in some sense "called forth" or "triggered" in response to demands for the satisfaction of certain classes of "needs." At the same time, after the buoyant optimism of the early 1960s, disillusionment over the utility of pure research has set in; in America, Federal funding of basic research at continued high levels has been challenged as not contributing to the solution of pressing economic and social problems, and NSF has established its Research Applied To National Needs program. Clearly, these new attitudes reflect more than merely the influence of the empirical studies, but such studies have seemed to provide ample justification for new policies that emphasize the decisive importance of "demand-pull," and at the same time minimize the potential effectiveness of "supply-push" policies.

Yet, whatever the merits of this questioning of the earlier assumptions and goals of science and technology policies, the notion that market demand forces "govern" the innovation process is simply not demonstrated by the empirical analyses that have claimed to support that conclusion. This paper will provide a critical survey of a number of the empirical studies of innovation, in order that their findings and methodologies, as well as their general conclusions, may be evaluated. We will show that the concept of demand utilized in many of these studies is a very vague one, often so broad as to embrace virtually all possible determinants of the innovative process, and therefore to rule out almost all other influences. In addition, as we will see, the studies really do not all examine the same dependent variables, but rather focus upon a number of different aspects of the innovation process in varied environments. This realization has escaped the attention of many writers who have based particular policy recommendations upon the "evidence" of these studies.

Our purpose, obviously, is not to deny that market demand plays an indispensable role in the development of successful innovations. Rather, we contend that the role of demand has been overextended and misrepresented, with serious possible consequences for our understanding of the innovative process and of appropriate government policy alternatives to foster innovation. Both the underlying, evolving knowledge base of science and technology, as well as the structure of market demand, play central roles in innovation in an interactive fashion, and neglect of either is bound to lead to faulty conclusions and policies.

What is needed to sort out the issues and improve the formulation of policy is a stronger predictive theory of innovation, one that goes beyond simply considering the motivations of individual firms to undertake research projects, and that deals instead with the mechanism by which this motivation may be transmuted into an innovative response of a particular sort, one influenced heavily by such factors as the supply of applicable science and technology (factors that also influence the nature of the problems posed for an innovative solution in the first place). Rather than simply referring to "lags" in the process, a useful theory of innovation must try to explain the varied length and distribution of such delays in the response to "needs." What factors serve to focus the problem-solving search upon certain solutions and when do these solutions appear? Such an analytic schema must explicitly consider institutional structures and dynamics, rather than the static analyses to which many theorists are wed, with some precision; the analysis should also be applicable at a fairly high level of aggregation, allowing international, interindustry, and intersectoral comparisons to be made.

We will first examine the empirical studies, followed by a discussion of how the "findings" of these studies are being interpreted and extended. In the concluding section, we discuss some policy implications of the critique.

Myers and Marquis

One of the most important and frequently-cited of the studies purporting to demonstrate the primacy of demand in the innovative process is that by Myers and Marquis (1969). The aim of their study was "to provide empirical knowledge about the factors which stimulate or advance the application in the civilian economy of scientific and technological findings" (Myers and Marquis, 1969, p. iii), focusing particularly upon the development by private firms of design concepts and the character and sources of information used in the subsequent development of innovations. Based upon an examination of some 567 innovations in five different industries, Myers and Marquis conclude that "Recognition of demand is a more frequent factor in innovation than recognition of technical potential" (Myers and Marquis, 1969, p. 60).

However, Myers and Marquis use an extremely loose definition of "market demand" in reaching this conclusion, and their taxonomy of the innovative process does not support such a sweeping statement.

The basic methodology employed in the study was that of interviewing executive and technical personnel in various firms within each of five industries; railroads, railroad-equipment suppliers, housing suppliers (e.g., paint and glass producers), computer manufacturers, and computer suppliers (e.g., peripheral equipment). These industries were chosen "to permit comparison of more technologically sophisticated industries with those which are less technically advanced" (Myers and Marquis, 1969, p. 10). However, the basis for classifying firms as members of these industries is nowhere specified, and, with the exception of railroads, none of the industrial categories correspond to those employed in the Census of Manufactures.[3] Rather than dealing with different general classes of innovation in their study, the authors are actually concerned with innovation in the producer-goods industries only, as are most of the other studies examined here. This group of industries may be characterized by a greater degree of "consumer sovereignty" than other industrial sectors, as the qualities demanded in products are communicated more clearly to producers in this sector than is the case for consumer goods industries.[4] It is important to remember throughout this survey that the studies discussed are concerned with a subset of the range of product and process innovations, and "lessons" from producer goods innovations should be applied elsewhere only with considerable care.

The innovations examined by Myers and Marquis were chosen by executives in 121 firms in the five industries in response to a questionnaire asking for the "most important" innovations in the previous five

[3] Thus, railroads are lumped together with four manufacturing industries. In addition, the composition of the housing, railroad, equipment, and computer suppliers' industries is quite troublesome, since firms in all of these industries are often very diversified. Housing products comprised less than 10 percent of the output of some firms placed by Myers and Marquis in the housing supplies industry.

[4] In the terminology of the studies by Teubal and the Falk Innovation Project, discussed below, "market determinateness" is quite high in these industries. Pavitt and Walker have pointed out that "users of innovations in industry and utilities in general know what they want, and whether they are getting it; when they are not, they can say so, which means that innovating firms aspiring to be successful must be sensitive to user needs. However, the same is not necessarily true in consumer markets or in those for public services, which are very far from 'perfect' in the sense described above . . . the market for public services is often fragmented . . . neither public services, nor individual consumers, may have the information and technical competence (or the resources to acquire it), necessary for the specification or evaluation of desired innovations" (Pavitt and Walker, 1976, p. 43).

to ten years. The innovations examined are thus all successful, and highly successful ones at that – a selection bias that, it is argued below, the Myers and Marquis study shares with others, and one that influences the conclusions reached in such analyses. The authors also asked the executives to assess the cost of these innovations; but no systematic assessment of the differential importance of individual innovations was undertaken. Such an index, whether defined in terms of increased sales or pecuniary savings, could have been used to weight the various innovations by their importance, a procedure that might have produced rather different results.

In conducting the personal interview with company executives and technical personnel, the authors employed a two-stage taxonomy of the innovation process, consisting of "(1) the idea-formulation stage, in which the recognition of potential demand and of technical feasibility fuse in a design concept, and (2) the problem solving stage leading to a testable item" (Myers and Marquis, 1969, p. 40). Prior to these two phases in the innovation process, according to the authors, is that of "recognition of both technical feasibility and demand" (Myers and Marquis, 1969, p. 3). In their schema, "demand" can be either current demand or potential demand, which largely deprives the concept of market demand of any operational meaning. Potential demands may exist for almost anything under the sun, and the mere fact that an innovation finds a market can scarcely be used as evidence of the undisputed primacy of "potential demand-pull" in explaining innovation. Nowhere do Myers and Marquis deal with the existence and size of any lags between these three phases of recognition, conceptualization, and problem-solving: their framework thus has little predictive power, in that it does not specify any speed of response with which demands "call forth" innovations. Nor, since we are dealing only with a universe of successful innovations, can we examine situations where demand-pull has failed to generate a successful response. All that has really been established is that there was an adequate demand for those innovations that turned out to be successful. We agree, but how would we disagree?

The authors' intent is to focus upon the conditions underlying the production of an innovation, and not at all upon the conditions surrounding its diffusion (commercial adoption). They exclude the phase of "implementation" from consideration, meaning by this term the actual production or use by the innovating firm of the product or process. However, the fact that only highly successful innovations are analyzed implies the existence of an identification problem, as it is difficult in retrospect to separate factors which underlay idea formulation and problem-solving from those influencing diffusion.

The authors conclude from their extensive interviews concerning the stimuli to undertake the research activity leading to a successful innovation that the "primary factor in only 21 percent of the successful innovations was the recognition of a technical opportunity. Market factors were reported as the primary factor in 45 percent of the innovations and manufacturing factors in 30 percent, indicating that three-fourths of the innovations could be classed as responses to demand recognition" (Myers and Marquis, 1969, p. 31). This is a very important finding that has since been given wide currency in the literature, and its empirical and conceptual underpinnings therefore merit close examination. What are considered to be instances of "technical opportunity"? These are defined as "Cases in which the dominant and immediately motivating factor was the perception of a technical opportunity to create or improve a product or the production process" (Myers and Marquis, 1969, p. 79). This is the most restrictive definition of "technology push" possible, one in which a technical breakthrough somehow appears serendipitously, and is applied without any consideration of commercial feasibility. Obviously, *both* commercial and technical feasibility are crucial to successful innovation; the blind application of technical or scientific knowledge implicit in Myers and Marquis's definition of "technological opportunity" is dysfunctional for the capitalist entrepreneur, to say the least.

The conclusions of the study concerning the stimuli for innovation, moreover, are drawn from a heterogeneous set of cases; 23 percent of the 567 innovations examined by the authors in this section of the study are in fact not innovations developed by the firm interviewed, but rather were adopted from some external source. In considering the factors motivating an individual firm to devote resources to research and development, the influence of the underlying scientific and technological knowledge base may be fundamental (we contend that this influence is vastly underestimated in these studies). The decision to adopt an externally developed innovation, however, is not one that is so dependent upon the constellation of technical opportunities that may arise out of the existing stock of scientific and technical knowledge. The innovation, after all, *already exists*. It is therefore inevitable that the inclusion of this substantial subset of firm-level decisions, which involve adoption rather than innovation, will lead to an underestimation of the "push" to innovation afforded by the underlying knowledge base.

The definitions of "market" and "production-related" influences upon the idea generation process, and the lumping together of these two categories under the rubric of market demand in Myers and Marquis, are open to serious question.[5] Of the three categories of influences upon the innovation process, none can be said to be unambiguously "demand-pull" according to any rigorous definition of this concept. Changes in performance or reliability standards are not necessarily identical with a shift in the demand curve. The use of "anticipated potential demand" is very suspect, particularly since it is a concept that really can have meaning only in hindsight, and additionally because of its weakness as a predictive concept; myriads of very deeply felt needs exist in the world, any one of which constitutes a potential market for some product, yet only a small subset of these potential demands are fulfilled. The recognition of the potential demand may precede the innovation by centuries; does this length of time constitute a mere "lag" in the innovation process?

Data on the length of time required for the metamorphosis of a "design concept" into a successful innovation are nowhere provided by Myers and Marquis; without some such information, no predictive or disprovable theory of innovation can be developed from these findings. The final "market-related" category, "response to competitive product," is basically a defensive response theory of innovation and product differentiation, denoting a rather different theory of innovation from that considered to be "demand-pull." For here, the innovation is "called forth" not by an increase, but rather by a *shrinkage*, in demand for a specific firm's product, due to the actions of a competing firm. This hardly constitutes support of a demand-pull hypothesis – to put it mildly – since the innovation is elicited by an anticipated *reduction* in demand! One might expect this final category of "recognition" to produce the least fundamental or radical innovations; in the absence of any data from the authors, we do not know. This defensive-response pattern of innovation, if it exists, also may lead to an over-investment in R&D by individual firms, for each of whom the private return exceeds the social return to innovative activity (see Hirschleifer, 1971).

The "production-related" categories pose some equally difficult problems of definition; most of them seem to point toward a theory of

[5] "Market-related" influences include "changed market requirements" (11.3 percent of the innovations), defined as "Cases in which the firm encountered a changed market for its product, either one requiring high performance or reliability standards or a change in demand"; "Anticipated demand" (23 percent), which are "Cases in which firm undertook the innovation because it anticipated an opportunity to market a 'new' product," or more of a modified existing product; and finally, "Response to competitive product," (10.7 percent), which are "Cases in which a competitor's marketing of a new or changed product induced the firm to reexamine its own product" (Myers and Marquis, 1969, pp. 78–79).

innovation that emphasizes technical interrelatedness or complementarities far more than market demand as the fundamental influences. These "production" factors include "Change in production process or design which made the innovation or related change necessary" (5.6 percent); "Change in production process or design which made feasible the innovation or related change" (5.3 percent); "Quality failure or deterioration" (1.1 percent); "Attention drawn to high cost" (6.2 percent); "Attention drawn to problem or inefficiency" (10.2 percent); and "Purchase of new equipment, including replacements of old" (0.5 percent) (Myers and Marquis, 1969, p. 79). All of these factors are doubtless of importance in focusing problem-solving attention and thus influencing the direction of innovation. It is very difficult, however, to see how these factors can be considered as primarily market demand-related, since they do not in any obvious way deal with either actual or anticipated increases in market demand for the products of the firms surveyed. Indeed, the influences classed as "production-related" by Myers and Marquis are not mediated by the market at all, but are internal to the firm.

"Production-related" factors could affect market demand in a very specific set of circumstances. If the market demand curve faced by a firm manufacturing capital goods is shifted by a change in the relative prices of the factors of production used by customer firms, one may posit a connection between the influences that we classify as production-related or supply side ones, and market demand. However, this more complex linkage is not set forth in any of the studies reviewed here. Such a case, in which a shift in relative factor prices alters a firm's choice of technique and thereby influences the direction of technical change, is one that has long been set forth as inducing a factor-saving bias in the direction of technical change. The proposition that changes in the relative prices of factors of production affect the direction of innovation, however, is by no means a proven one; debate over this issue has filled the historical and theoretical literature in economics.[6] Furthermore, while the choice of technique employed in production by a firm can change, this may affect only the adoption and diffusion of a particular innovation, not the appearance of a specific type of innovation. To make any predictive statement about the timing and specific character of an innovation, one must know more than the fact that a demand exists for innovation.

Myers and Marquis also collected data concerning the "problem-solving," final design phase of the innovation process. These data pri-

[6] Important works in the historical debate include Rothbarth (1946), Habakkuk (1962), and Temin (1966). Theoretical considerations of the issue are in Hicks (1932), Fellner (1962), and David (1975), ch. 1.

marily dealt with the sources and characteristics of the information that influenced the final, specific character of the new product or process. Again, one is struck by the artificiality of the authors' conceptual approach to innovation. How can the processes, and therefore the influences upon these processes, of "recognition" and "problem-solving," be meaningfully distinguished? Is not the innovation process entirely one of problem-solving, of loosely directed search? The Myers and Marquis vision of a complete "design concept," emerging out of an ill-specified phase of "recognition," seems grossly oversimplified. Adherence to this schema causes the authors to misinterpret their own empirical results, particularly in the analysis of the informational inputs to the problem-solving phase of the innovation process.

The authors categorize the impact of information utilized by the innovating firm either as having "evoked the basic idea," or having "expedited the solution" of the problem.[7] Their findings on the influence of technical or scientific information in the innovation process seem to contradict their previous conclusion that 75 percent of the innovations studied were responses to market demand. For if the timing, specific form and design, indeed function, of an innovation are direct results of the technical or scientific information utilized in its development, how can one also claim market demand as the dominant influence upon the rate and direction of innovation? In the analysis of information inputs to the innovation process, we find that 21 percent of the innovations categorized in the earlier analysis of motivational factors as responses to demand were classified in this section as innovations in which the problem was not even being worked on, until some technical or scientific informational input provided the basis for an idea. It is very difficult to see any significant difference between this class of innovations and innovations that were motivated by "technical factors."

In their examination of the role of scientific and technical information in innovation, the authors find that the evocation by some informational input of the basic idea for a problem that was not being studied or worked on accounts for 27 percent of the innovations examined in

[7] The first category, comprising some 27 percent of the cases considered, was defined as "cases in which the information provided the basis for the idea (or 'point of departure') for an innovation. The innovation may be viewed as solving a problem which previously had not been conceptualized." Information "expedited the solution" in the other 73 percent of the innovations considered by Myers and Marquis. In 55.9 percent of the cases, the expediting information "stimulated the idea or provided the solution of the problem being worked on"; in 9.2 percent of the innovations, the information simply "expedited the progress of the innovation being worked on"; information "narrowed the area of solution or demonstrated whether or not feasible" in 7.9 percent of the innovations (Myers and Marquis, 1969, pp. 80–81).

this section of the study. In addition, ideas or solutions for 55.9 percent of the innovations were stimulated by technical and scientific information; that is, the innovations probably would not have appeared without some informational input. In all, then, some 82.9 percent of the innovations considered by Myers and Marquis were heavily (or totally) dependent upon some technical or scientific information for their appearance (Myers and Marquis, 1969, pp. 80–1). Again, if one is to develop a predictive framework to understand innovation, explanation of the form and timing of innovations is rather important. Can one accept the conclusions of the authors about the importance of market demand if in fact some 83 percent of their innovations are so heavily dependent for their emergence upon the stock of technical and scientific knowledge? Clearly, Myers and Marquis have not provided a persuasive empirical case to support their belief in the "primacy" of demand forces in that innovative process.

Langrish et al.

The work by Langrish, Gibbons, Evans and Jevons, *Wealth from Knowledge* (1972), is one of the most painstaking of all of these empirical analyses of innovation. This group undertook detailed case studies of some 84 innovations that received the Queen's Award for technological innovation in 1966 and 1967, and all of which were commercially successful. The objectives of the study were

> first to provide some rather detailed accounts of what actually happened in recent instances of technological innovation . . . we aimed in particular to relate the technological to the organizational and other aspects. [Langrish et al., 1972, pp. 4–5]

The methodology of this study was broadly similar to many of the others reviewed here, although particular care was taken to ensure compatibility among the case studies and to gain a substantial familiarity with the technical and commercial context of each of the 84 innovations, nearly all of which were product innovations of a producer-goods nature. An initial request for general information from one of the innovating firms was followed by an intensive survey of the scientific, technical, commercial, and patent literature relevant to the particular innovation. The objectives of this analysis were to gain an appreciation for the environment out of which the innovation arose and the technical, commercial, and competitive conditions that surrounded the development process. Only after this survey were the technical and managerial personnel in each firm interviewed. The interviews were conducted with an eye to obtaining nonconflicting answers to similar questions from personnel at different positions in the firm; as the authors noted, "questions about the origin of an innovation led to quite different answers and this led to the realization that in many cases, even from the point of view of one firm, innovation is not a simple linear process with a clearly defined starting point" (Langrish et al., 1972, p. 84).

For 51 of the 84 innovations, the "major technical ideas or concepts" utilized were identified (158 ideas or concepts in all) in an effort to study the sources of, and transmission channels for, knowledge utilized in the innovation process. No attempt was made to categorize the functional or organizational origins of these ideas, however, as was done in the HINDSIGHT and TRACES studies. Although no analyses of the economic impact of the innovations were undertaken, a very important feature of this study is the attempt to classify the innovations by the size of the change in the technology that each represented, defining a particular technology as "a body of knowledge or industrial practice . . . sufficiently developed to provide a university course at M.Sc. or final-year B.Sc. level" (Langrish et al., 1972, p. 65). The method by which this change was measured was rather crude, consisting of ranking each innovation on a five-point scale according to the revisions in a standard course textbook necessary to accommodate the innovations.

Although this study's conclusions, as we will see, are frequently cited in support of the primacy of market demand in the innovation process, the actual approach adopted in *Wealth from Knowledge* was one that rejected the demand-pull schema in its cruder forms, as the authors concluded that "linear models" of the innovation process were unrealistic, noting that "the sources of innovation are multiple." This rejection of the linear model of innovation derived in part from the conceptualization of scientific and technical research as separate areas of endeavor, with only limited interaction. What little interaction takes place was viewed by the authors as a two-way flow of ideas and resources, rather than a undirectional flow from science to technology. In discussing the importance of "need," which is not well distinguished from market demand, the authors noted that

> Clear definition of need plus efficient planning fails to account satisfactorily for the majority of innovations. It is not exceptional for the need eventually met to be different from that foreseen and aimed at earlier on. [Langrish et al., 1972, p. 50]

The authors concluded that "perhaps the highest-level generalization that it is safe to make about technological innovation is that it must involve synthesis of some kind of need with some kind of technical possibility" (Langrish et al., 1972, p. 57).

A number of statistical analyses of the data for all 84 innovations were carried out by the authors. Of all the innovations examined, only 11 (13.1 percent) were fairly major changes; the separate analyses of this class reveal some interesting contrasts. Seven factors were identified as playing an important role in all of the innovations. Of these seven factors, "need identification" was found to be important in 16.7 percent of all innovations: "This factor is, or course, present to some extent in most innovations, but in some cases it was possible to demonstrate that the identification of a need was a major reason why the Award-winning firm succeeded in the innovation instead of its competitors" (Langrish et al., pp. 67–8). "Need identification" was the second most prevalent factor identified as important in the innovations examined.[8] Here, as in the other studies, the failure to include commercially unsuccessful innovations means that some of these factors must be viewed as aiding in the commercial diffusion and adoption of the innovations, rather than influencing their invention and/or development processes. In addition, some of the factors such as resource availability, or good intrafirm cooperation, surely must have been important in most if not all of the innovations, and perhaps should be excluded. "Need" was nowhere defined with any precision and, as such, cannot be considered to be identical to market demand.

The study proceeded to consider separately the major and minor innovations, yielding some interesting results. For major innovations, the recognition of a discovery's usefulness was found to be far more important than was true in the total sample (occurring in 14.4 percent of the cases of major technological innovations, the third most prevalent), while "need identification" accounted for only 6.1 percent (tied for fourth place). The relative importance of the two factors was reversed in the minor innovations, as the need factor occurred in 18.3 percent and discovery-push was of importance in only 5 percent of the cases, a finding similar to that in the Battelle study's analysis of research events (see below).

Some evidence of a weakly counterfactual variety concerning the factors that distinguish successful and unsuccessful innovations is contained in the analysis of factors that delayed innovation: "These catego-

[8] The six other factors, and the percentage of innovations examined in which they were found to be of importance are: "Top person: the presence of an outstanding person in a position of authority" (25.1 percent); presence of some other indispensable person, who was of crucial importance due to his or her specialized knowledge (14.7 percent); "Realization of the potential usefulness of a discovery" (6.2 percent); "good cooperation" (4.9 percent); "availability of resources" (8.2 percent); and "Help from governmental sources, including research associations and public corporations" (5.3 percent), (Langrish et al., 1972, p. 69).

ries refer to delays between an innovation being apparently possible and the Award-winning firm achieving success" (Langrish et al., 1972, p. 70). Success was nowhere defined, but it apparently denotes successful commercial introduction of an innovation. As such, some of these factors refer to the diffusion process.

This analysis of delaying factors is of importance in providing the basis for a more powerful predictive theory of innovation, in that it focuses upon the period between the inauguration of organized research and the commercial introduction of an innovation. The extremely loose treatment of this period in many empirical analyses of innovation, which refer to it as a "lag," renders such analyses distinctly suspect as foundations for policy and prediction. The authors found that the most important factor delaying successful innovation, occurring in 32.5 percent of the cases, was the insufficient development of some other technology: "In many cases an innovation was possible but could not be successfully developed until another technology had 'caught up' " (Langrish et al., 1972, p. 70). Thus, supply-side factors, that is, technological complementarities, are the most important variables in accounting for the timing of innovations. In 22.5 percent of the cases, there existed at first "no market or need," and the potential of the innovation was not recognized by management in 7.6 percent of the cases. Interestingly, when the major and minor innovations were analyzed separately, the "lack of market" and "lack of complementary technology" factors were of equal importance for major innovations, while for minor innovations, the lack of complementary technologies was more important than the lack of a market. These findings for all of the innovations do little to support the hypothesis that market demand governs the innovation process; were demand the most important influence, one would expect its absence to be of primary importance in explaining delays or failures, which is not the case here.

Four varieties of "linear models" were considered by the authors in a discussion of the explanatory power of simple models for the 84 innovations examined. Under the rubric "discovery push" are placed two models, the first of which, the "science discovers, technology applies" model, is defined as one "in which innovation is seen as the process whereby scientific discoveries are turned into commercial products." The alternative discovery push model is the "technological discovery" one, for "Many innovations are not clearly based on any scientific discovery but can be described as being based on an invention or technological discovery" (Langrish et al., 1972, p. 73). The first of two heuristic models introduced under the "need pull" category is the "customer need" model, in which the innovation process is described as one "which starts with the realization of a market need. Market research or a direct request for a new product from a customer can be the start of research and development activity." The "management by objective" model was apparently intended as one dealing with more general "mission-oriented" innovation efforts: "Some innovations can be described in terms of the start of the process being a need identified by the management where this need is not a customer need. For example, the need to reduce the costs of a manufacturing process can lead to resources being allocated to research and development" (Langrish et al., 1972, p. 74). Of course, the "need" on the part of a firm to reduce its own manufacturing costs should be classified as a supply-side and not a demand-side phenomenon, inasmuch as it has little to do with the market demand for the products of a given firm. While need is still not distinguished from demand, the distinction between customer and internal needs is a useful one, the absence of which in many other analyses of factors underlying successful innovations is unfortunate.

Having constructed these heuristic models, the authors conclude that very few of the innovations "fit any of the above models in a clear and unambiguous manner. The reason for this is quite simple. It is extremely difficult to describe the majority of the cases in terms of a linear sequence with a clearly defined starting point" (Langrish et al., 1972, p. 71). This statement, with which we are in complete agreement, renders curious the frequent citation of this study as evidence supporting demand-pull theories of innovation (see below).

In focusing upon the factors that stimulated the activity leading to the innovations, a more precise formulation of the Myers and Marquis "design concept" phase, the authors provided some support for the role of demand (whether this implies anything about the nature of the innovation process is quite another issue; focusing upon the motivating factors alone represents a retreat to "black box" theorizing). Even at this remove, however, the authors noted that "it is still very difficult in a large number of cases to state clearly that the innovation is of one type or another" (Langrish et al., 1972, p. 75). When combinations of the two models in each category were utilized, however, "need pull" was roughly twice as frequent as "discovery push" in the firms' decision to initiate the activities resulting in the innovation. On the other hand, it is significant that for major technological changes the "discovery push" models accounted for seven of the eleven changes so categorized.

It is important to note the restrictiveness of this finding. It is not concerned directly with the factors underlying successful innovation (nor do the authors claim this), but merely the initiation of the search activities leading to the innovation. Since we have no evidence regarding the motivating factors for a sample of unsuccessful innovations, the implications of this finding are not clear; nor, of course, do findings about the motivation for R&D tell one much about the specific characteristics of the outputs from the innovation process. Yet one of the coauthors recently referred to the study as follows: "A study of 84 British technological innovations showed that 'demand pull' occurred more frequently than 'discovery push' " (Langrish, 1974, p. 614). This is a rather strained interpretation of the findings of *Wealth from Knowledge;* it was not "demand" but "need" that was pulling, and the "pull" referred to dealt in fact only with the motivation to commit resources to research and development, not at all with the influences upon the rate and direction of outputs.

HINDSIGHT

The Department of Defense study of the contributions of research to the development of weapons systems, HINDSIGHT,[9] is another empirical analysis often quoted in support of the "demand pull" theories of innovation. However, the study does not consider commercial innovations, so the reliance upon its conclusions in support of the demand-pull hypothesis requires the further logical step of equating market demand with Department of Defense "needs"; this point will be discussed further below (certainly, the difficulties of such defense contractors as General Dynamics in selling high-technology products in civilian markets suggests that there are some considerable hurdles to be made in the logical step). In addition, the methodology of HINDSIGHT exhibits failings very similar to those of the other studies reviewed here: A linear model of innovation is imposed retrospectively upon a very complex interactive process; basic and applied research activities are conceptualized as competitive, rather than complementary; and the technological events are arbitrarily assigned identical weights.

In the postwar period, the Department of Defense supported a great deal of research in universities and the private sector; in 1967, some $400 million went to research (out of a $7 billion R&D budget), of which $100 million was basic, "undirected" research and $300 million was "applied" in nature. According to its two directors, HINDSIGHT was intended both to "measure the payoff to Defense of its own investments in science and technology" and "to see whether there were some patterns of management that led more frequently than others to usable

[9] Office of the Director of Defense Research (1969); see also Sherwin and Isenson (1967) and Greenberg (1966).

results . . . In particular we wanted to determine the relative contributions of the defense and nondefense sectors" (Sherwin and Isenson, 1967, p. 1571). The development of 20 weapons systems was analyzed by Department of Defense scientists and engineers for HINDSIGHT. The basic methodology was that of benefit-cost analysis; the costs of performing a given military function with the current, recently developed weapons system were compared with the costs of utilizing the predecessor system, at an equivalent level of effectiveness. The pecuniary savings were attributed directly to the research expenditures incurred in systems development and were found to be substantial. The study concluded that "the approximately $10 billion of Department of Defense funds expended in the support of science and technology over the period 1946–1962 . . . has been paid back many times over" (Sherwin and Isenson, 1967, p. 1575).

In trying to address the second set of issues, those dealing with project management, HINDSIGHT attempted to identify important discoveries or breakthroughs that made possible successful development of the weapons systems. Seven hundred ten such "Research Events," each consisting of "the occurrence of a novel idea and a subsequent period of activity during which the idea is examined or tested" (Office of the Director of Defense Research and Engineering, 1969, p. 139), were identified and classified as Science or Technology Events, according to the evaluators' perceptions of the intentions with which the work leading up to each Event was being carried out.[10]

All Events were weighted identically in assessing the development of these weapons systems, and HINDSIGHT's conclusions regarding the relative importance of mission-oriented versus basic Research Events in weapons systems development have aroused considerable discussion. Only 9 percent of the Events deemed as essential to the development of the weapons systems were found to be Science Events, while 91 percent were Technology Events. Even more important was the finding that of the 9 percent of all Events dubbed Science Events, 97 percent were motivated by a need, only 3 percent arising from "undirected" scientific

[10] "Science Events" included mathematical and theoretical studies dealing with natural phenomena, "experimental validation of theory and accumulation of data concerning natural phenomena," and a combination of the above two activities. "Technology Events" included the development of new materials, "Conception and/or demonstration of the capability to perform a specific elementary function, using new or untried concepts, principles, techniques, materials, etc.," and theoretical analyses of properties of materials or equipment (Office of the Director of Defense Research and Engineering, 1969, p. 12). The study noted that the definition of "scientific" research was a very rigorous one, and was in fact more restrictive than that employed by the Defense Department (Office of the Director of Defense Research and Engineering, 1969, p. 60).

research. Twenty-seven percent of all of the Events were directed at "a broad class of defense needs not related to a single system or system concept"; 41 percent of all Events were "motivated by a system or system concept in the early or 'advanced development' stage"; and 20 percent of all Events were Technology Events dealing with systems in more advanced stages of development.

In a passage later to be misinterpreted as evidence for the primacy of market demand forces in the innovation process, HINDSIGHT's directors concluded that "nearly 95 percent of all Events were directed toward a Department of Defense need" (Sherwin and Isenson, 1967, p. 1574). Assuming away the numerous conceptual and methodological difficulties of HINDSIGHT, can this evidence concerning the 20 weapons systems and their development be applied to support the *market* demand-pull hypothesis? In the first place, it is far from clear how this evidence can possibly be related to innovation in a market context; are such authors enthusiastically quoting HINDSIGHT viewing each of 710 Events as individual innovations, or are each of the 20 weapons systems seen as the unit of analysis, the innovations, that were "called forth" by a need? Nowhere is this specified, but certainly individual Research Events are not anything like innovations in a capitalist market context (nor does HINDSIGHT claim this for them); rather, they are intermediate steps in solving the larger problem of building a weapons system, the desired characteristics of which are well specified by Department of Defense personnel. To conclude with HINDSIGHT that certain well-specified technical problems in electronics and metallurgy can be solved by research personnel is a distinctly different conclusion from that which portrays market demand as governing the direction and character of innovation.[11]

The view of innovation that is implicit in the study's attempt to assess the direct contributions of basic and applied scientific and technical research is one that seems to view scientific and technical research as somehow being competitive, rather than complementary, enterprises. Yet, such a view of the innovative process is a very peculiar one, at variance with most accepted "linear" models of innovations. In these formulations of the innovation process, scientific research serves as a foundation for technical research – certainly this is a prominent justification for the Pentagon's funding of university basic research. It is likely that "linear" models of innovation greatly exaggerate the extent to which the flow of ideas and resources from basic to applied research is

[11] See Peck and Scherer (1962), ch. 3, entitled "The Nonmarket Character of the Weapons Acquisition Process."

unidirectional in nature. Nonetheless, the basic research contribution to more applied technical work, in terms both of personnel and information, is crucial. Indeed, in the executive summary of the study's Final Report, a passage apparently overlooked by many of the interpreters of the findings of HINDSIGHT states:

> It is emphasized that this study identified only those incremental contributions to existing bodies of scientific and technological knowledge that were utilized in the analyzed military equipments. The strong dependence of these contributions upon the total base of science and technology must be recognized. [Office of the Director of Defense Research and Engineering, 1969, p. xv]

Operationally, the resort by HINDSIGHT to an arbitrary cutoff point of 20 years in tracing the important Research Events in weapons development works to underestimate the contribution of basic research activities. Clearly, much of the knowledge base that is absolutely essential to the technical breakthroughs in electronics, for example, goes back to the late nineteenth and early twentieth centuries, as the Final Report points out:

> less than 16 percent of the technologically oriented R&D Events were traced to a post-1935 science base. The other 84 percent came directly from the application of nineteenth-century unified theory, were the results of empirical research, or appeared as invention not needing scientific explanation. [Office of the Director of Defense Research and Engineering, 1969, p. 31]

The findings of the study thus do not address the relative importance of basic versus applied research activities, but rather concern the vintage of the scientific knowledge that is utilized in the weapons innovation process. The importance of the basic scientific knowledge base is underlined by the above statement, which also reveals the sensitivity of the study's findings to the definition of the time horizon for Research Events, and the care with which these findings must be interpreted.

The notion in HINDSIGHT that breakthroughs that culminated in a new weapons system can simply be "added up" overlooks the great variation in the importance of individual Research Events. This flaw is not unique to HINDSIGHT, but it serves to bias the findings against the importance of Science Events, given the often-greater importance of basic research findings, due to their wide applicability and radical nature. The HINDSIGHT schema also views each innovation as a logical outcome of an accumulation of technical breakthroughs consciously directed toward a single end; yet this is belied by a glance at many of the case studies compiled by the project. In many cases, significant findings resulted from work on a completely unrelated project; yet the study takes all "mission-oriented" events as directed toward the particular weapons system under examination in the case study. The approach also ignores the great uncertainties inherent in the innovation process, the numerous dead ends and unsuccessful Research Events – what proportion of all of the mission-oriented research projects failed, and how does this compare with the basic scientific research projects? Nowhere does HINDSIGHT address these issues – again, we have a study dealing only with successful innovations, in this case, successful "Research Events."

TRACES and BATTELLE

TRACES (Technology in Retrospect and Critical Events in Science) (Gibbons and Gummett, 1977) and the Battelle Research Institute study on the *Interactions of Science and Technology in the Innovative Process* (Battelle Research Institute, 1973) are broadly similar analyses of a few innovations, adopting a lengthier time horizon and focusing more upon the early phases of the innovation process than was the case in HINDSIGHT. These studies share with HINDSIGHT a concern with the contributions to innovation of basic and applied research, as well as a methodology of examining identifiable "events" in the innovation process (indeed, many observers have portrayed these studies, commissioned by the National Science Foundation, as responses to HINDSIGHT). The use of a lengthier time horizon in these studies yields a more reasonable view of basic and applied research activities as complements, rather than substitutes, in the innovation process. The Battelle study represents a more ambitious effort to analyze the relative importance of research events, as well as the factors influencing these events; while TRACES made no reference at all to the relative importance of "demand-pull" or "technology-push" in the innovation process, the Battelle study did attempt to analyze such factors.

TRACES is a collection of reconstruction of the key research events involved in each of five innovations: magnetic ferrites, the video tape recorder, oral contraceptives, the electron microscope, and matrix isolation. The study was intended to "provide more specific information on the role of various mechanisms, institutions, and types of R&D activity required for successful innovation" (Illinois Institute of Technology, 1968, p. ii).

The retrospective tracing of the innovation events (not, it should be noted, the underlying *processes*) for each of the five innovations was performed by scientific and technical experts; only the most important events, defined as "the point at which a published paper, presentation, or reference to the research is made" (Illinois Institute of Technology, 1968, p. 2), were included. Once identified, research events were classified on the basis of their "technical content or motivation," in a fashion similar to the procedure followed in HINDSIGHT. No attempt was made to weight differentially the research events according to their importance, which again is a technical limitation shared with HINDSIGHT. Unlike the HINDSIGHT study, no distinction was drawn between "scientific" and "technological" research events and activities. Three categories of general R&D activity were considered in the study: nonmission research, "motivated by the search for knowledge and scientific understanding without regard for its application"; mission-oriented research, "performed to develop information for a specific application concept prior to development of a prototype product or engineering design"; and development and application, "involving prototype development and engineering directed toward the demonstration of a specific product or process for purposes of marketing" (Illinois Institute of Technology, 1968, p. ix).

As we have seen in the case of HINDSIGHT, the definition of the time horizon for each innovation examined is crucial in this "reconstructive" methodology. The time horizon adopted in TRACES is one that places far more emphasis upon the development of the scientific knowledge base, and much less upon the phase of modification and refinement for commercial sale. The definition of the origins of an innovation is extremely elastic, but it does exclude the diffusion process from the history of a given innovation.

The conclusions of the TRACES study present an interesting (if not surprising) contrast to those of HINDSIGHT. Some 341 research events were identified; of these, some 70 percent were found to be of the "nonmission-oriented" variety, while 20 percent were mission-oriented, and 10 percent were in the development category. The contrast with the conclusions of HINDSIGHT is due in part to the finding of TRACES that 45 percent of the nonmission research underlying an innovation was completed at least 30 years in advance of the innovation, and thus excluded *a priori* from consideration in HINDSIGHT, and the study's findings that 80 percent of nonmission research is completed 15 years before the innovation.

TRACES dealt primarily with the contributions of basic or applied research to innovation, the central issue raised in HINDSIGHT. The Battelle study, on the other hand, was commissioned by NSF as an extension of TRACES, and was designed to deal with a broader range of issues (including those considered by Myers and Marquis). In addition, an effort was made in the Battelle study to rank research events by importance. The study reexamined three of the five innovations that had been analyzed by TRACES (magnetic ferrites, oral contraceptives, and the video tape recorder), and considered an additional five: heart pacemakers, hybrid grains, xerography, input–output economic analysis, and organophosphoric insecticides. Research events were identified by scientific and technical personnel as before, with the additional technique of oral interviews being utilized extensively. As in TRACES, the three categories of nonmission, mission-oriented, and development events were utilized in the categorization.

The important differences between TRACES and the Battelle study arise in three areas. As was mentioned above, the Battelle study attempted to distinguish the subset "decisive" events from "significant" events in the innovation processes considered. Significant events are essentially identical, in definition and dating, to the key events of TRACES, a significant event being defined as "an occurrence judged by the investigator to encapsulate an important activity in the history of an innovation or its further improvement, as reported in publications, presentations, or references to research." A decisive event "provides a major and essential impetus to the innovation. It often occurs at the convergence of several streams of activity. In judging an event to be decisive, one should be convinced that, without it, the innovation would not have occurred, or would have been seriously delayed" (Battelle Research Institute, 1973, p. 22).

A second major difference is the analysis contained in the Battelle study of the influence of some 21 factors upon decisive events, as well as an attempt to define certain characteristics of the overall process of innovation in each of the eight cases considered. Among the 21 factors are Myers and Marquis's "motivational factors," *viz.*, "recognition of scientific opportunity," "recognition of technical opportunity," and "recognition of the need," as well as managerial, external, and peer group influences. Nowhere did the Battelle study specifically identify the ways in which these various categories influence decisive events – in their analysis of these motivational factors, Myers and Marquis were speaking only of the motivation for a single firm to undertake a particular R&D project, rather than of factors motivating specific scientific or technical discoveries. In addition, the Battelle study utilized a chronological framework that weighted the final development phase more heavily; this difference with TRACES is not mentioned explicitly but emerges from a comparison of the chronologies in the two studies for similar innovations. The sensitivity of the findings of each study to these different approaches underlines forcefully the difficulty of finding objective truths in this area.

A total of 533 significant events were identified in the Battelle study, of which 89 (17 percent) were judged to be decisive. Of the 21 factors adduced as influential in the innovation process, "recognition of technical opportunity" and "recognition of need" were the most important for the decisive events, occurring for 89 percent and 69 percent, respectively (managerial and funding influences followed these in importance). The finding concerning need recognition for these individual events is difficult to interpret, but it does not seem to support the "market demand-pull" hypothesis, concerned as this study is with individual events in the overall process rather than the motives inducing a single firm to undertake a particular R&D project. The authors of the Battelle study assessed their findings rather differently, however: "Other studies of innovation have shown that innovations are most frequently responsive to the force of 'market pull,' that is, to a recognized consumer need of demand. This observation is broadly confirmed in the present study, even though the ratings cover individual decisive events, rather than the innovation as a whole" (Battelle Research Institute, 1973, p. 3-1). If need recognition plays such an important role in structuring individual events in the innovation process, this probably reflects the importance of technical complementarities acting as focusing devices, rather than anything like the importance of a market demand for a research breakthrough. The research breakthrough *per se* is not generally marketed, but contributes to an innovation that may be sold commercially.

Problems also arise in the Battelle study's overall characterization of the case studies for each of the innovations. Again, there is nowhere a clear specification of where or how these characteristics enter the process. The study concluded that "in nine of the ten innovations studied (hybrid grains were counted as three innovations: corn, wheat, and rice) recognition of the need for the innovation preceded the availability of means to satisfy that need. This finding supports several previous studies of the innovative process" (Battelle Research Institute, 1973, p. 4-6). However, this conclusion is as empty of meaning as many of those reached by the other studies. Can one seriously maintain that the "need" for high-yield grains, oral contraceptives, or a heart pacemaker has been unrecognized for the past centuries? In any case, of what possible explanatory relevance is the temporal priority of the "recognition of the need for the innovation?"

The Battelle study's findings regarding the importance of the three categories of research in producing significant and decisive events differ somewhat from those of TRACES, reflecting the arbitrary beginning date and the slightly different emphasis in each of the retrospective case studies. Nonmission-oriented research accounted for 34 percent, mission-oriented research was responsible for 38 percent, and development for 26 percent of the significant events. Moreover, when the analysis was confined to the decisive events, the study concludes that only 15 percent arose out of nonmission-oriented research, compared with 45 percent and 39 percent, respectively, for mission-oriented and development activities:

> Presumably, this fact reflects the tendency for decisive events to occur within or close to the innovative period. Events that occurred at times long preceding the innovation period might have occurred some years later without delay to the date of first conception or first realization of the innovation. Such events, often in the realm of NMOR (nonmission-oriented research) therefore cannot be considered, in retrospect, decisive at the time of the innovation. [Battelle Research Institute, 1973, p. 4–8]

Gibbons and Johnston

It is worth introducing here the results of a different approach to the retrospective evaluation of innovation and the contributions to it of basic and applied research. Gibbons and Johnston (1974) undertook an analysis of thirty innovations that focused upon the origins and character of informational inputs used to solve technical problems in the innovation process. This approach to analyzing the contributions of in-house or external research, as well as basic and applied research, conceptualizes technical problem-solving as based largely upon the exchange of information. The authors maintained that their study "has avoided any attempt to trace the origins of a piece of hardware to either science or technology" (Gibbons and Johnston, 1974, p. 230). This contrasts to the methodology of the HINDSIGHT, TRACES, and Battelle studies.

Gibbons and Johnston classified information inputs as scientific or technical by the source, rather than the content, of such information. This poses at least one possible problem, inasmuch as purely technical information may be transmitted through papers published in scientific rather than technical journals. An additional methodological difficulty exists in the authors' efforts to establish the age of the information inputs used in the innovations. If the lag between discovery and publication is significant and varies among different kinds of knowledge, analyzing the timing of the application of scientific work to innovation through examining the published work is likely to yield deceptive results.

Nevertheless, the conclusions of Gibbons and Johnston are potentially very significant in improving our understanding of the complex role of science in the innovation process. They found that scientific information made a significant contribution to the innovations. Thirty-six percent of externally obtained information, and 20 percent of all information, utilized in the innovation processes examined was scientific in character; the information of greatest utility in solving technical problems was drawn from scientific journals. Personal contacts with scientists from outside the firms (usually from universities or government research establishments) were found to yield very useful information, not only about theories and properties of materials, but of a more directly useful, "applied" variety, concerning the availability of specialized equipment or suggesting alternate designs. In many cases, university scientists aided firms as a sort of "gatekeeper," " 'translating' information from scientific journals into a form meaningful to the 'problem-solver' " (Gibbons and Johnston, 1974, p. 236). University-educated personnel within the firm were able to contribute substantially to the problem-solving process due to their greater use of external scientific information sources, in the form of journals or acquaintances at universities.

Gibbons and Johnston concluded from their analysis of information flows that the interactions between basic and applied, in-house and external, research are so complex that normative criteria for the optimal direction and size of government research support are not easily discerned. They also concluded that the basic research infrastructure in universities and governmental installations contributes to commercial innovation in ways other than simply providing the private firm with exploitable scientific discoveries. Thus a complex, nonlinear relationship between basic and applied research is indicated by Gibbons and Johnston's work, far more than is the case in HINDSIGHT. Freeing the analysis of innovations of the restrictions imposed by a Procrustean focus upon "events" thus allows for recognition of a more complex interrelationship between innovation and basic research.

Carter and Williams

C. F. Carter and B. R. Williams (1957, 1959) carried out an extensive study of innovation decisions of British firms for the Board of Trade, in an effort to illuminate the factors that aid or hinder the application to industrial products and processes of scientific research. As the subtitle of *Industry and Technical Progress* suggests, the basic conceptual schema in their first study is very much a linear model of the innovation process. The discussion in *Industry and Technical Progress* is concerned with the stages in the generation and application of scientific knowledge, from basic research to the point of substantial investment in the commercial production of an innovation, as well as the managerial and economic factors that facilitate the application of such knowledge on the part of private firms. The central concern of *Investment in Innovation,* published subsequently, is the specific factors influencing the decisions of firms to invest in new technologies. One major difficulty in the analyses carried out by the authors is that of distinguishing the factors governing the diffusion of an existing innovation from those influencing the development of a new one.

The methodology utilized in the Carter and Williams studies was that of extensive field interviews and case studies. In-depth studies of 152 firms were carried out through interviews and questionnaires, covering a wide range of firm characteristics, including information on executives and research personnel, firm size and product range, and communications patterns within firms. The knowledge base underlying each firm's industry was also reviewed to gain an appreciation of the potential for innovation within each industry. In addition, a small sample of firms in each of four industries – paper, pottery, jute, and cutlery – and equipment suppliers in each was examined in order to provide a more intensive study of the innovation process. Five specific innovations in pottery, building, metals, and machinery were also studied to provide an understanding of the transition from invention to diffusion.

The authors were reluctant to impose a highly structured model of the innovation process upon their data; rather, they chose simply to report the results of their extensive surveys. The authors concluded that explicit consideration of market demand factors is rare in decisions concerning the level of investment in research and development. This was due to an absence of clearly defined corporate objectives and a focus upon short-term development and survival. Yet one certainly would expect consideration of market demand to be of great importance in the firm's allocation of resources to research and development if the market demand factors are crucial in affecting the direction of innovation. The findings, therefore, are not supportive of the demand-pull hypothesis. Evidence of a slightly different nature, concerning the origins of specific R&D projects, is presented by the authors for slightly less than one-third of 200 innovations examined. They found that roughly 25 percent of the innovations originated in a firm's R&D laboratory, 18 percent began from the prediction that a future market existed for a particular product, and 10 percent of the innovations each were contributed by the demands of sales departments and production departments for solutions for immediate problems. Such evidence of course deals only with the inception of research and, additionally, the forecast "need for an innovation," strictly speaking, does not imply an increased market demand. Profit-seeking firms are most unlikely ever to proceed in the *absence* of some anticipated "need."

Evidence concerning the conditions influencing the commercial success of an invention is found in both *Industry and Technical Progress* and *Investment in Innovation*. The precise nature of the dependent variable in these surveys is not made very clear; is it diffusion, or simply the development of an invention to the point of commercial feasibility? In *Industry and Technical Progress*, the firm's decisions to proceed with development were examined in 250 cases:

> In sixty-five cases (fifty-three involving the introduction of a new product differing markedly from the normal range) the firms were induced to proceed by the conviction that the potential market was large. In fifty-four cases improvement in quality of product was thought to improve the competitive position of the firm. In sixty-nine cases the expectation of a saving in process cost was a sufficient motive. In thirty-five cases the fact that there was a "significant innovation" was thought to be in itself a sufficient reason for going ahead. In eighteen cases the innovation, by broadening the basis of the firm's operations, was thought to improve the prospects of long-term stability; in sixteen cases innovations were intended to guarantee supplies of materials; in ten cases "government interest" made the project seem worthwhile. In thirteen cases the fact that the innovation "increased output" was a sufficient incentive. [Carter and Williams, 1957, p. 85]

Here one finds that "potential demand" was the primary factor in only slightly more than one-quarter of the cases, even under the fairly elastic definition set forth.

Investment in Innovation examined the investment decisions of business firms that underlay the diffusion of new technologies. In a sample of 204 innovations, "just over a half . . . were ascribed by the firms concerned to definite causes, such as the pressure of competition or of excess demand" (Carter and Williams, 1959, p. 57). Of these, 18 percent were adopted out of a desire to overcome shortages of materials or labor; "desire to meet excess demand" accounted for 10 percent; "demands by customers for new types or qualities of product" was responsible for 12 percent; "desire to use the work of research and development departments" occurred in 18 percent; "direct pressure of competition" accounted for 10 percent; and "successful trials of product or process in other industries or in other countries" explained 33 percent of the events. Finally, the authors distinguished between "active" and "passive" innovative decisions:

> In making a *passive* innovation the firm is responding to direct market pressure, as shown by excess demand or by developing competition and falling profit margins . . . In making an *active* innovation the firm is deliberately searching for new markets and techniques, even though there may be no direct market pressure to do so. [Carter and Williams, 1969, pp. 57–8]

It seems clear that only a subset of the passive innovations developed by a firm can be viewed as "called forth" by market demand. Carter and Williams's finding that 60 percent of the 204 innovations considered are of the active variety is therefore of interest.

The Carter and Williams study exhibits few of the methodological problems of the other studies considered here; this is largely due to the lack of a well-specified analytic framework and the (perhaps understandable) reluctance to draw strong general conclusions from the study. Particularly troublesome is the recurrent fuzziness about precisely what aspect of the innovation process is being considered in a given set of surveys. In only one set of survey results, the study of the departmental origins of research and development projects, do we find support for the primacy of market demand forces in the innovation process. This study provides some rather weak evidence supporting the view that market or customer needs serve to influence the direction of enterepreneurs' innovative efforts. We certainly do not deny this. It is the identification of "needs" with "market demand," and the dominant role in commercial innovation ascribed to this amorphous variable, that we criticize. As has been noted elsewhere, however, evidence concerning the *origins* of R&D projects is of limited relevance to attempts to understand the rate and direction of innovation. Much of the other evidence set forth by the authors either does not deal with the innovation process *per se,* or does not afford any support for a demand-pull position, or both.

Baker et al.

The studies by Baker, Siegman, and colleagues (1967, 1971) provide perhaps the most dubious collection of evidence concerning influences on the innovation process. The first paper, by far the more widely quoted of the two, is concerned with the more general phenomenon of idea generation within organizations, using a corporate research laboratory as the object of study: "we are concerned with identifying organizational factors which aid or deter idea creation, and with describing how a set of ideas was created in one large industrial R&D laboratory" (Baker et al., 1967, p. 157). One of the authors defined an idea as "an actual or potential proposal for undertaking new technical work which will require the commitment of significant organizational resources" (Baker et al., 1967, p. 158). The object of study is thus only tangentially related to the innovation process; all that is considered is the very preliminary phase of project proposal. The decision as to which ideas merit further development, the direction and character of the subsequent research process, and the commercial introduction of an innovation are all excluded from the purview of this article.

The authors attended meetings of "idea generation groups" (IGGs) at a corporate research laboratory, in the role of silent observers and note-takers. The ideas put forth by the research group were rated on a numerical scale, and the 5 to 10 "best" ideas submitted by the group were chosen by the participants in the idea generation group. Nowhere are the criteria, upon which these ratings were based, made explicit – nor is any effort made to distinguish subsequently successful ideas from unsuccessful ones. A set of "stimulating events" (e.g., "thinking by self," "interaction with other persons") was identified and ideas classified accordingly; in addition, "the nature of the problems which stimulated the ideas" (Baker et al., 1967, p. 159) was considered. "Need events" were defined as "recognition of an organizational need, problem or opportunity" and were identified for 94 percent of the ideas developed; "means events," defined as "recognition of a means or technique by which to satisfy the need, solve the problem, or capitalize on the opportunity," occurred for 92 percent of the ideas. Two "patterns of idea-generation behavior" were also identified, "need–means" and "means–need," referring to the sequence of the above two events in the generation of ideas. Seventy-five percent of the ideas, and 85 percent of those subjectively rated as the best ideas, were categorized as arising from "need–means" sequences, while 25 percent resulted from the converse sequence.

In a subsequent article, "The Relationship between Certain Characteristics of Industrial Research Proposals and Their Subsequent Disposition," the authors extended their analysis of idea generation to include the phase in which a choice among the ideas produced by the IGG is made. The authors found that urgency, defined as "Immediacy of the need, problem, or opportunity toward which the idea is directed" (Baker et al., 1971, p. 119), was positively correlated with a subsequent affirmative decision to proceed with the research project. Again, however, "need" is very broadly defined, and the decision to proceed with research tells one little about the successful or unsuccessful development of an innovation.

Interpreting the findings of these studies as supportive of the "market demand-pull" hypothesis seems very strained; although others have done so, the authors of the articles made no such claims. Their definition of "needs" is extremely broad and overlaps with the concept of market demand to only a limited extent. Taken together, the articles do address the nature of the stimuli to undertake research; this represents a small part of the overall innovation process, however, and the subsequent technical and/or commercial success or failure of the innovation is not considered in such a framework.

SAPPHO and FIP

A final pair of studies is noteworthy for their attempts to compare directly successes and failures in innovation. Both SAPPHO (Scientific Activity Predictor from Patterns with Heuristic Origins), from the Science Policy Research Unit at the University of Sussex, and FIP (Falk Innovation Project), from the Falk Institute for Economic Research in Jerusalem, compare innovations that failed either before (i.e., research is terminated) or following introduction.[12] The success criterion thus deals more with commercial diffusion than with the innovation process *per se*. In interpreting the conclusions of the studies, therefore, it is important to note that their central subject is not the factors that motivate the processes leading to successful or unsuccessful innovations, nor is it the character of the innovation process itself. In addition, the fact that both studies focus upon innovation within one or two industries implies that the scientific and technical knowledge base upon which the firms under examination may draw is fairly *homogeneous,* and "technology-push" factors are therefore likely to be excluded *a priori.*

The two studies differ in that "while SAPPHO is basically a study of the management of innovation, FIP is an attempt at identifying comparative advantage in innovation" (Rothwell et al., 1976, p. 2). Thus, while SAPPHO performed pairwise comparisons of commercially successful and unsuccessful innovations within two industries (chemicals and scientific instruments), with an eye to identifying characteristics differentiating the members of each pair, FIP attempted to focus upon features of market and producers' structures that would explain the pattern of success and failure for a whole range of innovations within a single industry (electronic medical instruments). The FIP study thus subsumes considerations of project selection, as well as project manage-

[12] R. Rothwell, C. Freeman, A. Horsley, V. T. P. Jervis, A. B. Robertson, and J. Townsend (1974 [SAPPHO]); M. Teubal, N. Arnon, and M. Trachtenberg (1976 [FIP]); and R. Rothwell, J. Townsend, M. Teubal, and P. T. Speller (1976 [SAPPHO/FIP]).

ment. Both studies contain some extremely useful analyses of the influence of marketing factors upon commercial successes in innovation, and as such represent important refinements in the theory of demand-induced innovation.

Forty-three pairs of innovations in chemicals and scientific instruments were analyzed in the course of SAPPHO, drawing upon extensive interviews and technical analyses. In the first phase of the study, the analysis was confined to assessing the importance of 24 variables in a strictly pairwise manner, comparing commercially successful innovations with unsuccessful ones. No broader ranking of all of the innovations examined, whether in terms of their importance or degree of success, was undertaken. Subsequently, an analysis of groups of innovations was carried out in an effort to focus upon the variables influencing the degree of commercial success. Five categories of variables were found to be most important in distinguishing successes from failures:

(1) Successful innovators were seen to have a much better understanding of user needs.
(2) Successful innovators paid more attention to marketing and publicity.
(3) Successful innovators performed their development work more efficiently than failures but not necessarily more quickly.
(4) Successful innovators made more use of outside technology and scientific advice, not necessarily in general but in the specific area concerned.
(5) The responsible individuals behind the successful innovations were usually more senior and had greater authority than their counterparts in unsuccessful projects (Rothwell et al., 1974, pp. 285–6).

Some interesting contrasts between the chemicals and instrument industries also emerged in the explanation of success and failure. The chemicals innovations were predominantly process innovations developed by very large firms; the influence of individuals was less important, as was the understanding of user needs, than was true in the instruments industry, characterized by small firms and product innovations.[13] Lead times, that is, being the first with a new product or process, were quite impor-

[13] A recent series of papers by von Hippel concerning innovation in the American scientific instruments industry emphasizes the active role of instrument users as originators of innovations. Von Hippel's work underlines the importance of "user needs" in successful product innovation, but his studies do not support a "market demand pull" hypothesis for innovation. Instead, they tend to highlight the importance of the distinction between "needs," which may arise from any source, and for any reason, and market demand, which is perforce mediated through a market. The "needs" to which innovative instrument users are responding are in very few cases equivalent to consumer demands transmitted through the market. See von Hippel (1976, 1977).

tant in chemical industry innovations, which frequently were quite radical changes, while in instruments, the first firm to introduce an innovation failed more often than did the second innovator. These interindustry differences are a central aspect of the FIP and SAPPHO/FIP studies.

A brief discussion in SAPPHO of the motivation for innovations concluded that "Of the 43 pairs studied, in 21 percent 'need-pull' differentiated in favor of success: in the remaining 79 percent there was no correlation between success and failure and motivation: in no case did technology-push differentiate in favor of success, or need-pull in favor of failure" (Rothwell et al., 1974, p. 277). The definitions of "need-pull" and "technology push" are not reproduced in the article. It is again worth noting the similarity of technologies in most of the pairs of innovations studied, a methodological aspect that may have biased the results of the analyses against "technology-push" explanations.

The FIP study did not proceed with a strict pairwise comparison of innovations, but rather focused upon the performance of R&D programs of nine firms in a single industry, biomedical electronics. Among the variables examined were so-called area variables, characteristics of firm structure and market demand in the industry. This set of influences was not examined in SAPPHO; however, most of the other variables considered in the analysis of success and failure in the FIP study were analogous to those utilized in SAPPHO. Also similar to SAPPHO was the definition of success, which was simply commercial success. The FIP study examined the number of successful and unsuccessful research projects within each of these nine R&D programs; despite the rejection of the SAPPHO pairwise approach in favor of a more comprehensive, intraindustry analysis, no ranking of the extent of success or failure was attempted.

The study concluded, as did SAPPHO, that sensitivity to user needs was of great importance in explaining the success of research projects. One piece of evidence adduced in support of this position stated that "The proportion of failures in programs whose idea originated in R&D exceeds the proportion of failures in other programs" (Teubal et al., 1976, p. 18), which obviously requires an assumption about the relative sensitivity of a firm's various units to "user needs." The same caveats concerning the definition of success, and similarity of underlying technologies, that were noted for SAPPHO, apply here as well. Of particular interest in the FIP study's discussion of market demand, however, is the concept of "market determinateness," one of the exogenous variables in the study: "Market determinateness refers to the degree of specificity of the market signals received by the innovating firm and consequently to the extent to which it anticipates (instead of responding to) demand" (Teubal et al., 1976, p. 20). The similarity between the firm behavior implied by this definition, and the Carter and Williams concept of active and passive innovative firms is interesting and may reinforce the contention that only the passive firms really fit the "demand-pull" paradigm. In all of the extensive discussions of the role of demand in the innovative process that have been surveyed here, this concept stands out as a useful one for comparing the role and importance of "demand-pull" forces in different industries, something that cannot be done merely using the broad, fuzzy concepts of "user needs" or "need recognition." Comparing all of the R&D programs within the biomedical electronics industry, the FIP study concluded that "The degree of market determinateness differentiates sharply between the successful and unsuccessful biomedical electronics R&D programs studies" (Teubal et al., 1976, p. 31).

In a comparison of the FIP and SAPPHO studies, it was concluded that an important interindustry explanatory variable lay in this area of "market determinateness":

> Most of the SAPPHO chemical innovations were process innovations not involving fundamental changes in existing products while most instrument innovations involved a significant new-product component. This suggests that the chemical innovations were more market determinate than the instrument innovations, which may in turn explain the greater effort required to understand user needs and the correspondingly greater need for external communications in the instrument industry. [Rothwell et al., 1976, p. 11]

The analysis of innovation failure thus yields rather useful insights, particularly this last concept of market determinateness, a concept that affords a means of examining and comparing innovational performance at a somewhat higher level of aggregation than do most of the "demand-pull" approaches to the problem. However, neither SAPPHO nor FIP focuses narrowly upon successful innovation *per se*, but includes in the definition of success much that more appropriately should be seen as relating to the diffusion of innovations; the support afforded by these studies for "market demand-pull" explanations of innovation is thus distinctly limited.

Interpretations of the studies

A major reason for surveying the empirical studies of innovation is the widespread acceptance of the conclusions or, at least, the frequent appeals to the findings of these studies in support of "demand-pull" positions. As was pointed out repeatedly in the review above, most of these studies say rather little about the influence of market demand upon the rate and direction of innovative output. However, even the limited conclusions of the studies are frequently distorted and misquoted. A sampling of the secondary and review literature is discussed below to show the extent of the haphazard reliance upon these studies. Much of the problem derives from the fact that the quoted studies differ widely in the nature of the dependent variable that is being explained, and often deal with very different phases of the innovation process.

Myers and Marquis's empirical study contains a brief summary of four other studies that, they argued, supported their conclusion that "demand recognition" is the driving force in the innovation process. Three of the studies, Carter and Williams, HINDSIGHT, and the first article by Baker et al., were reviewed above. The authors compare their findings on the influences that motivate firms to undertake R&D projects with the findings of Carter and Williams, published in *Investment in Innovation* (not *Industry and Technical Progress,* incorrectly referred to by Myers and Marquis as the source), concerning a much broader range of investment behavior. As was noted above, a central focus of *Investment in Innovation* is the influences upon the diffusion process and adoption decisions of firms, not simply the factors motivating research; these latter factors are discussed in *Industry and Technical Progress.*

Myers and Marquis also discussed the Carter and Williams categories of "active" and "passive" innovating firms, without stating in what fashion these findings of *Investment in Innovation* support the "demand-pull" position.[14] As was discussed above, however, the empirical results of the active/passive distinction do not strongly support the primacy of "demand-pull" influences upon innovation. The authors cited the first study by Baker et al. as support for their position, despite the fact that the Baker study does not even deal with commercial innovation, but rather with the very different subject of idea generation. The interpretation of HINDSIGHT set forth by Myers and Marquis also ignores the fact that the HINDSIGHT study examines Research Events, rather than innovations, and the objects of study have little to do with innovation in a market context.

A recent review article by J. M. Utterback in *Science* draws upon eight empirical studies of innovation, five of which were reviewed above,[15] in stating that

[14] "Active innovating firms produce most of the innovation studied" (Myers and Marquis, 1969, p. 33).

[15] Utterback (1974). The five studies reviewed above to which Utterback refers are Baker et al. (1971), Langrish et al. (1972), Carter and Williams (1957), Rothwell et al. (1974), and Office of the Director of Defense Research and Engineering (1969).

> market factors appear to be the primary influence on innovation. From 60 to 80 percent of important innovations in a large number of fields have been in response to market demands and needs . . . There is a striking similarity between the findings of studies conducted in the United States and those conducted in the United Kingdom. [Utterback, 1974, p. 621]

SAPPHO, however, really deals with the conditions underlying commercial success of innovations, while the Langrish study provides some support for "need-pull" only as a motivating influence and rejects simplistic, linear models of innovation explicitly. Utterback nowhere made clear what specific aspect of the innovation process he was addressing, an omission of importance because of the differing foci of the empirical studies that he quotes. In addition, the equation of "market factors," "market demands," and "needs," implies an unacceptably loose definition of market demand and "demand-pull." The problems of using HINDSIGHT and the Baker study as support for Utterback's contention have already been discussed.

Carter and Williams's *Industry and Technical Progress* is referred to by Utterback as having shown 73 percent of 137 innovations to have been stimulated by "market, mission, or production needs," while 27 percent arose from "technical opportunities." In fact, nowhere in *Industry and Technical Progress* does there appear such a result for an analysis of 137 innovations. However, the authors' examination of "the passage from invention to innovation in over 250 cases of product or process development" concludes that 188 instances, 75 percent, resulted from the "expectation of a saving in process cost," "the conviction that the potential market was large," or improvement in the "competitive position of the firm" (Carter and Williams, 1957, p. 85). Again, the connection of these factors with shifts in market demand is far from obvious and the phase of the innovation process discussed is not that of the initial decision to proceed with R&D. Myers and Marquis's study and that by Langrish et al. are also relied upon for empirical support by Utterback. The conclusions of both studies concern the motivation for innovation, although in both studies, as we have seen, the distinction between need and market demand never is drawn sufficiently sharply to provide unambiguous support for Utterback's position.

A recent study of the evidence on *Technology, Economic Growth and International Competitiveness* by Robert Gilpin (1975) contains a strong endorsement of the "market demand-pull" approach to innovation as a basis for government policy "to stimulate the technological innovations and industrial productivity required to help meet international economic competition, stimulate economic growth, and solve our domestic problems" (Gilpin, 1975, p. 1). Although Gilpin's position apparently relies upon the evidence from the studies reviewed here, he does not quote any of them, preferring instead the sweeping assertion that

> Everything that we know about technological innovation points to the fact that user or market demand is the primary determinant of successful innovation. What is important is what consumers or producers need or want rather than the availability of technological options. Technological advance may be the necessary condition for technological innovation and on occasion new technology may create its own demand but in general and in the short-run, the sufficient condition for success is the structure or nature of demand. [Gilpin, 1975, p. 65]

This statement is an extreme version of "demand-pull," and in fact contradicts Gilpin's own earlier statement that "successful innovation involves increasingly a *coupling* or *matching* of new science and technology with market demand" (Gilpin, 1975, p. 37), a statement implying that demand is necessary, but not sufficient. This confusion of necessity and sufficiency is really at the heart of much of the controversy over the role of demand in innovation. Gilpin further strikingly qualifies the statement by admitting that "This emphasis upon market demand as a stimulant to innovation must be qualified with respect to innovation in one area of economic goods, that is consumer goods and services" (Gilpin, 1975, p. 38). In other words, "everything we know about technological innovation" in fact relates to a rather specialized subset of economic sectors, the more so since military and space industries cannot be said to exist in a conventional market environment. Needless to say, this is rather a substantial "qualification"! As if this were not sufficient qualification to his sweeping assertions about the primacy of market demand, Gilpin also concedes that radically new technologies can and do create their own markets. "Certainly this has been the case with such radical innovations as the computer, the laser, and nuclear power" (Gilpin, 1975, p. 37). One is reminded of Schumpeter's observation about Malthus: that the most interesting aspects of his theory of population were the qualifications.

The only piece of empirical work on the role of demand-side forces in innovation that Gilpin explicitly discusses is *Invention and Economic Growth* by Jacob Schmookler: "Schmookler demonstrated that the primary factor in successful innovation was market demand" (Gilpin, 1975, p. 37). This statement represents an illegitimate extension of Schmookler's findings, and reflects the confusion and ambiguity over the precise nature of the dependent variable mentioned above. Schmookler's work deals with invention, *not* commercially successful innovations; thus his use of patent statistics as a measure of inventive output. Rather than explaining the factors underlying commercially successful innovations, Schmookler analyzed market demand forces as they influenced shifts in the allocation of resources to inventive activity – an entirely different matter. Gilpin is at once more and less careful in his interpretation of the evidence than others discussed here. While he refers only to market demand factors in his discussion, which presumably is the restrictive set of influences dealt with by economists, Gilpin nowhere makes clear the fact that these many studies of the innovation process rarely refer to narrowly defined market demand as the crucial motivation.

Conclusion

A general critique and discussion of the empirical studies reviewed above is undertaken in this section. The studies may be criticized at the specific level of failing to substantiate their hypotheses; the primacy of market demand forces within the innovation process is simply not demonstrated. At a more general level, however, the weaknesses of the broad conceptual framework of the studies become clear; the uncritical appeal to market demand as the governing influence in the innovation process simply does not yield useful insights into the complexities of that process. A brief discussion of some possible policy implications of these issues concludes the paper.

A major difficulty exists in the interpretation of the results of these empirical studies, in that they vary widely in the nature of the dependent variable being considered, and their conclusions, even assuming they were well supported, therefore do not supply uniform support for a demand-pull argument. Myers and Marquis, Baker et al. and Langrish et al. deal with the motivations for the allocation of R&D inputs by a single firm in their conclusions supporting demand-pull theories of innovation. TRACES, HINDSIGHT, and the Battelle study all focus upon the lengthy histories of specific innovations, each of which involves numerous individuals, commercial and noncommercial laboratories, and firms. The "demand-pull" conclusions of these studies concern events *within* the process resulting in a given innovation, in numerous different institutional settings, rather than a single firm's decisions concerning the commitment of resources to particular projects. In addition, HINDSIGHT deals with the innovation process in a nonmarket environment, one in which commercial feasibility and market demand are concepts shorn of their usual meaning. The conclusions of the SAPPHO and FIP studies deal primarily with the factors influencing the commercial success of an innovation following the production of a prototype, again approaching the phenomenon at the level of the individual firm.

In order to retain its analytic content, market demand must be clearly distinguished from the potentially limitless set of human needs. Demand, as expressed in and mediated through the marketplace, is a precise concept, denoting a systematic relationship between prices and quantities, one devolving from the constellation of consumer preferences and incomes. To be taken seriously, demand-pull hypotheses must base themselves upon this precise concept, and not the rather shapeless and elusive notion of "needs." Yet most of these empirical studies rarely, if ever, make such a distinction, as a result of which the relationship of the "need recognition" category to market demand as a motivating or controlling influence in the innovation process is tenuous indeed. An additional consequence of this confounding of need and market demand is the frequent failure to distinguish between motivations or influences upon the innovation process that arise from within the economic unit, such as those resulting from increases in output or changes in production technology, from factors that are external to the firm and mediated by the market. Examples of this are the treatment by Myers and Marquis, as well as Utterback (in his interpretation of Carter and Williams's findings), of "production needs," which are internal to the firm, as equivalent to shifts in the market demand for a firm's products. Also of importance is the fact that all of the studies that attempted to rank innovations or research events by the importance of these occurrences found that the most radical or fundamental ones were those least responsive to "needs." Even within the flawed conceptual and methodological framework of these empirical studies, then, the "demand-pull" case is admittedly weakest for the most significant innovations. Does an explanatory schema retain much usefulness when it is contradicted by the most important occurrences in the set of events that it purports to illuminate?

The working definition of market demand employed in many of the empirical studies examining the individual firm's decisions in the innovation process is one that excludes only the most economically irrational set of decisions; that is, market demand becomes identified with *all* those price signals transmitted through the marketplace that provide the basis for rational economic decisions. The definitions of "technology-push" employed by Myers and Marquis imply that only cases in which absolutely no attention is paid to the economic return likely to result from an innovation can be considered as instances in which a "technological push" caused the innovation. Yet, in a capitalist economy, where decision-makers operate on the basis of expectations of

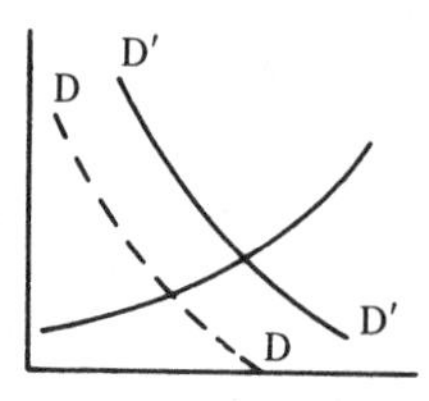

Figure 1. DD to D'D'.

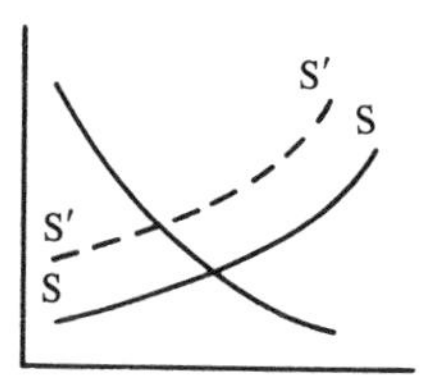

Figure 2. S'S' to SS.

future profit, no substantial innovation activity will be undertaken unless there is some reasonable expectation that there exists a market demand sufficiently large to justify that expenditure. In a technologically sophisticated capitalist society, with a range of versatile technological skills at the disposal of the capitalist decisionmaker, it would be sheer dereliction of responsibility were he or she *not* to consider the alternative courses of action available in terms of market size. At the same time, however, the range of actions that is available and their respective costs are being continually altered by the course of technical progress; the costs of alternative actions are always changing. This problem is exacerbated by the open-ended interview method used in these studies, in which businessmen are asked to reconstruct the decisions made in successful innovation processes. It seems obvious that no entrepreneur is going to admit to having gambled blindly on a technological potential alone, giving no thought at all to the profitability of its development. Nor is it realistic to expect an entrepreneur to plunge blindly ahead in such a fashion. Yet this in effect is the response necessary to demonstrate the primacy of "technology-push" in such a study. No wonder "technology-push" appears relatively unimportant!

In order to establish the proposition that market demand forces "called forth" an innovation, a shift in the demand curve must be shown to have occurred (Figure 4). This is not to be confused, of course, with a movement along the demand curve (Figure 5).

Distinguishing empirically between the types of parametric shifts raises an identification problem. Is an innovation introduced because the demand for a product has *increased* (i.e., the demand curve has shifted outward) or because technological improvements (or other sources of cost reduction) now make it possible to sell the product at a lower price (i.e., the downward shift in the supply curve leads to an intersection with the demand curve at a lower price than before)? The first case is the one required to support the "demand-pull" hypothesis. However, the information elicited from interviews with businessmen, given the nature and structure of the questions addressed to them, makes it virtually impossible to distinguish the demand-pull situation from that of technology-push. At the very least, as we have demonstrated, numerous instances of changes in production costs or requirements have been classified under "demand-pull."

When we ask why a particular innovation came at a particular point in time, it is never enough to say that it was "market demand." The question is why innovation did not come years earlier or later. The answer to such a question therefore has to deal with *changes* in demand or supply conditions. It is not sufficient to say that demand conditions "stimulated" or "triggered" an event; rather one must demonstrate changes in demand conditions. To establish the primacy of demand-side factors, one has to show that demand conditions changed in ways more significant or decisive than changes in supply conditions – for example, in cost. None of these studies deals with this identification problem.

The demand-pull approach reflects an insufficient appreciation for the innumerable ways in which changes, sometimes very small changes in production technology, are continually altering the potential costs of different lines of activity. It is these almost invisible, cumulative improvements in underlying technology, or the perception that it is possible to attain such improvements, or the realization that a cost-reducing technique in use in industry A is applicable (with perhaps minor modifications) in industries B, C and D, that play a critical and neglected role in accounting for the timing of innovative activity. The existence of an adequate demand for the eventual product is, of course, an essential – a necessary – condition. But, we suggest, the "demand-pull" approach simply ignores, or denies, the operation of a complex and diverse set of supply-side mechanisms that are continually altering the structure of production costs (as well as introducing entirely new products) and that are therefore fundamental to the explanation of the timing of the innovation process.

At a more general level, the conceptual underpinnings of the "demand-pull" case are perhaps even more fundamentally suspect. Rather than viewing either the existence of a market demand or the existence of a technological opportunity as each representing a sufficient condition for innovation to occur, one should consider them each as necessary, but not sufficient, for innovation to result; both must exist simultaneously. There is no good *a priori* reason in theory why "market demand" factors should be dominant in motivating innovative activity. If we assume the world to be populated by firms attempting to maximize profits, *any* environmental shift presenting an opportunity for technical change yielding a net profit will be taken, regardless of whether it derives from a shift in market demand, a change in the conditions of supply, or a promising technical breakthrough. A whole range of stimuli are important in the innovation process, not simply market demand. Moreover, any careful study of the history of an innovation is likely to reveal a characteristically iterative process in which *both* demand and supply forces are responded to. Thus, successful innovations typically undergo extensive modification in the development process in response to the perception of the requirements of the eventual users, on the one hand, and, on the other, in response to the requirements of the producer who is interested in producing the product at the lowest possible cost. Innovations that are not highly sensitive to *both* sets of forces are most unlikely to achieve the status of commercial success.[16]

The limited usefulness of the "demand-pull" conceptualization is demonstrated in attempts to apply the conclusions thence derived at a level of aggregation above that of the individual firm; there is no clear mechanism by means of which to move from the micro to the macro level of analysis with such a model. Thus, in comparing the innovative performance of nations, Pavitt notes that

> the empirical evidence suggests that there is in fact a weak relationship between the size and sophistication of national markets, and performance in technological innovation . . . much higher correlations with national innovative performance exist for "supply" rather than "demand" factors, such as the number of large firms, the level of industrial R and D, and capabilities in fundamental research. [Pavitt/OECD, 1971, p. 53]

This applies with equal force in the comparison of the innovative performance of industries within a national unit; whatever evidence exists (and little of an unambiguous variety may be said to exist) relating the rate of innovation to attributes of various industries in a cross-sectional comparison without exception deals with "supply-side" variables, including such factors as producer concentration and "technological opportunity."[17] The work of Schmookler concerning the cross-sectional comparison of industries uses patents as its dependent variable, and thus should be seen primarily as explaining the allocation of inventive effort, since the set of patented inventions bears a tenuous relation, as Schmookler recognized, to the much smaller set of commercially successful innovations. Not only do most patents never reach the stage of commercial exploitation, but many commercially successful innovations are unpatented. Moreover, it seems reasonable to ask of a theory of innovation that it allow one to compare the performance of

[16] This is shown in the SAPPHO (Rothwell et al., 1974) and FIP (Teubal et al., 1976) studies on the conditions responsible for commercial success in innovation, as well as in von Hippel (1977).

[17] See Scherer (1965) for a good empirical cross-sectional analysis.

industries and nations; this certainly is of importance for policy purposes. The "demand-pull" theories of innovation cannot be said to be useful theories on this score. These theories in no way provide a satisfactory account for one of the most distinctive features of industrial societies: the wide variations in the performance of individual industries with respect to observed rates of technological innovation and productivity growth.

In addition to providing very little useful information on the comparative performance of national and industrial aggregates in innovation, the "demand-pull" approach to innovation reveals little about the specific timing and direction of innovative *output*. Even if one grants, as we do not, a dominant role for demand in the innovation process as asserted in these empirical studies, their findings relate to the stimuli for innovation, the influences upon the allocation of R&D inputs, not the rate and direction of outputs, which are the variables that one wants to explain. Many of the findings of the studies also refer to attributes, or measures of the success, of innovations, rather than characteristics distinguishing success from failure in innovation. Clearly, an "understanding of user needs," or the "satisfaction of an anticipated demand," are really criteria that define successful innovations. They hardly qualify as independent explanatory variables. Many of these empirical studies, particularly those that analyze the process from the viewpoint of a single firm, adopt a "black box" explanation of the innovation process: inputs in, innovations out.

A recent article, "Problems in the Economist's Conceptualization of Technological Innovation" (Rosenberg, 1976, ch. 4), criticized the notion of a smooth, convex, well-defined isoquant in most microeconomic theorizing in which the structure of relative factor prices yields a determinate solution to the choice of technique problem under all circumstances. In many of these studies of innovation, the implicit idea of a well-defined innovation possibility frontier, based upon scientific knowledge, which underpins the black-box approach, is open to serious criticism on similar grounds: The search for the basic knowledge for solutions to problems is not costless, nor does the knowledge base merely fulfill a passive "blueprint" role in the innovation process.

The exclusive focus upon the outcomes of a complex process (in this case, innovations), which has long been the *modus operandi* of the economist, can be justified only in an environment of certainty and equilibrium, as Herbert Simon pointed out:

> The equilibrium behavior of a perfectly adapting organism depends only on its goal and its environment; it is otherwise completely independent of the internal properties of the organism . . . but when the

> complexity and instability of his environment becomes a central feature of the choices that economic man faces . . . the theory must describe him as something more than a featureless adaptive organism: it must incorporate at least some description of the processes and mechanisms through which the adaption takes place. [Simon, 1966, p. 3]

The notion that the market transmits clear and readily recognized signals for innovations is important to the validity of the "black box" approach to the analysis of innovation. Yet Nelson has noted that the existence of such a mechanism is a rather shaky assumption:

> The assumption of a well-perceived demand curve for product or supply curve for input is plausible only if one can describe mechanisms whereby these curves in fact get well perceived. This would seem to imply considerable experience on the part of the firms in the industry in the relevant environment of demand and supply conditions. This clearly cannot be assumed in an environment of rapid change in either demand or supply conditions. In particular, it seems completely implausible in considering the demand for a major innovation. [Nelson, 1972, p. 46]

Indeed, the pervasiveness of uncertainty in the innovation process is ignored by most of the empirical studies. This is particularly true of the studies that have as their unit of analysis some group of selected innovations, and that attempt to reconstruct the history of such innovations: HINDSIGHT, TRACES, and the Battelle study are prominent examples. The attempt to decompose neatly the complex, stochastic, and uncertain process which is that leading to innovations into a set of events that can simply be cumulated to yield an innovation, is as gross an oversimplification as is the "black box" approach of the surveys of business firms. To attempt such a reconstruction, and further to attempt to ascribe relative importance to the various categories of research events (as in HINDSIGHT), seems fallacious.

The innovation process surely comprises an area of economic behavior in which uncertainty and complexity are absolutely central characteristics of the environment; empirical approaches to the problem must therefore take far greater cognizance of the processes that underlie the output of innovations. Rather than focusing exclusively upon innovational outputs at widely separated times, a more fruitful approach might be that of tracing the growth and evolution of a given organizational form involved in the research and innovation processes, in an effort to provide a somewhat deeper analysis of the evolution of the information flows and processes that are responsible for success (or failure) in the production of innovations over a period of time; changing the unit of analysis in this manner might yield a richer set of conclusions and studies than is currently the case.

Some of the literature in the field of technological change produced by economists represents an effort to get at structural aspects of the innovation process, whether through the construction of innovation production functions or, more narrowly, by attempting to relate certain structural parameters of firms or industries to innovative performance or research intensity. However, this body of literature, for all its merits in explicitly examining the structural and institutional aspects of the process, is largely ahistorical. Empirical studies that pursue the historical background are confined to dealing with a particular innovation. Confining the unit of analysis to the single firm in these empirical studies also creates problems, due to the lack of attention devoted to interfirm relationships, that are often central in the understanding of technical change. This is clearly illustrated in an article by Abernathy and Townsend (1975) in which the unit of analysis is rendered more comformable with the underlying technology, yielding a substantially different set of conclusions dealing with the same data that Myers and Marquis considered. Just as much of the new work in industrial organization frequently concerns product lines, rather than the often – and increasingly – artificial boundaries of the firm, so should empirical analyses of technologies remove the blinders imposed by concentration upon the single firm in isolation. The present paper will have served a useful purpose, if only of a preliminary, "desk-clearing" nature, if it has persuaded the reader that the now widely accepted bit of conventional wisdom concerning the primacy of "demand-pull" forces in the innovation process is lacking in any persuasive empirical support.

Some policy implications

What implications follow from the discussion above for governmental science and technology policies? The fundamental conclusion is that the current stock of scientific and technical knowledge is *not* omnicompetent, and that an active role for government in affecting the rate and direction of innovation is necessary, one going beyond the prescription below:

> The emphasis of both direct and indirect government intervention in the economy should be to transform the market in ways which will encourage industry to innovate products of better quality and greater social utility . . . it should create the incentives and disincentives which will encourage industries to be more innovative in the use of their R and D resources. [Gilpin, 1975, p. 39]

While we agree with Gilpin's statement as far as it goes, it does not go nearly far enough. The government ought indeed to take whatever steps it can to improve the functioning of the private market and the complex incentive system that is mediated through market forces. But such measures, although eminently desirable, are not nearly sufficient. For the production of new knowledge that underlies and shapes the innovative process is, itself, very inadequately served by market forces and the incentives of the marketplace. The need for a more positive public policy is well recognized both by numerous scholarly studies that have documented the discrepancy between high social returns and low (or even negative) private returns with respect to investment in knowledge production, and the diverse and elaborate systems of public support of knowledge-producing activities that have emerged in all advanced industrial societies. Attempting to deal with such problem has, for example, produced quite different organizational arrangements in America with respect to such diverse sectors as medicine, aircraft, atomic energy, and agriculture. There are, in addition, many forms of knowledge that are urgently required for successful social policy, that will not be produced through any plausible system of private market incentives. Further thought concerning the most efficient ways of organizing the R&D process is urgently needed.

The point is that in certain areas, such as alternative energy or antipollution technologies, industry may simply lack sufficient R&D resources or the necessary market-generated incentives. In many industries and areas of substantial social need, we simply do not have the basic knowledge of scientific and technical phenomena to proceed intelligently; our limited understanding of such complex ecosystems as San Francisco Bay, for example, or of the effects upon human health of long-term exposure to certain industrial wastes, greatly hampers the development of optimal antipollutant technologies and regulations. It is important to understand that the record of postwar American technical dynamism is a direct outgrowth of scientific and technical research in a very few areas (such as electronics), often funded and justified by defense requirements. This knowledge is clearly transferable in certain cases – semiconductors are an obvious example – to the civilian sector, but it is limited in its range of applicability. Integrated circuits will not immediately eradicate urban blight.

The semiconductor electronics industry provides a useful example of an extremely progressive industry, which has been characterized by sensitivity to demand factors but which has also been fundamentally governed by the exploitation of a new body of scientific and technical knowledge.[18] The industry represents perhaps the most outstanding

[18] Both Golding (1971) and Tilton (1971), the two best references on the semiconductor industry, emphasize this point.

"success story," in terms of government policy to stimulate technical progressiveness, and growth in output and employment, in the postwar period in the United States. This revolution in electronics that has stretched over the last three decades does not represent a response to sudden shifts or increases in market demand, but rather reflects the increasingly wide exploitation of new technical capabilities, based upon advances in solid-state physics and production technologies. Clearly, the record of innovation in the industry has been one of sensitivity to the desires of customers (it is hard to see how matters could have been otherwise), particularly military customers. Yet, the case simply cannot be made that this demand for miniaturized solid-state components and logic functions was one that alone or primarily brought the industry into being. It is the exploitation of the radically new properties and capacities of semiconductors and integrated circuits that has been a mainspring of the industry's growth. Innovative effort has been directed at the solution of problems of application and production of this technology. The fundamental "focusing devices" have been internal to the development of the technology, rather than being explicable by reference to exogenous demand forces. Process innovations, grounded in theoretical and technical developments, have been fundamental in the history of product innovation in the industry.

An important role thus exists for government funding of certain kinds of scientific and technical research in a broader range of nondefense areas, perhaps including some limited involvement in the construction of pilot plants, similar to the DoD policy of the early 1950s concerning transistors. Attempting to revolutionize the technological underpinnings of certain sectors through the resort to a "big push" approach, however, is likely to fail – cognizance must be taken of the fact that the basic knowledge may be lacking, as well as the complexity of market demand and the interface between successful government and private development efforts. Of course, advocates of the "demand-pull" approach are absolutely right insofar as they insist that money poured into schemes for which no demand exists is likely to be money wasted. But, as we have seen, they claim far more than that for their position.

In addition to nourishing the supply side in a broader range of areas, intelligent policies must be directed at institutional aspects of the innovation process, working to encourage the interaction of users and producers, as well as the iterative interactions between more basic and applied research enterprises. We do not yet understand the characteristics of the innovation process sufficiently well, nor do we possess the necessary knowledge base in certain areas of substantial social utility. Useful policies would be those directed at the provision of information, from basic research institutions in the noncommercial sector to private firms and laboratories, as well as from users to producers concerning desired products and characteristics.

More generally, policies directed toward increasing both the frequency and the intimacy of interactions among these separate participant groups may prove to be particularly rewarding. This involves not only expanding the network on which information may flow among these groups but, additionally, measures which will increase the incentive to participate in such interactions.

An additional point of importance focuses upon the limited policy relevance of the "demand-pull" position, which was mentioned above. Based as it is upon the hypothesized motivations and actions of the individual firms, policies aimed at industries and sectors do not follow easily from the "demand-pull" conceptualization of technical innovation. A recent paper by Gibbons and Gummett discussed the efforts of the British government to develop structures to provide "demand-pulls" for the numerous research laboratories of the Department of Industry. The authors emphasize that it is still too early to assess the success of the new "Research Requirements Boards," but they note that the "problem of trying to aggregate individual strategies into sector strategies and, hence, sector needs has, to our knowledge, received very little attention so far . . . one is left in the dark about how these needs are to be identified" (Gibbons and Gummett, 1977, pp. 27–28). Such a set of difficulties inevitably will plague efforts to "call forth" commercially successful innovations from private or public research institutions.

It is important to point out in conclusion that the above policy prescriptions are directed at the encouragement of innovation, not the diffusion of innovations. This second problem is one of equal importance to the national policymaker concerned with productivity growth and related problems. Encouragement of the diffusion of innovations, however, seems to be an area in which one can indeed rely upon the more conventional market incentives designed to induce demand-pull for a proven innovation. Certainly, the policies proposed above that are directed at the improvement of communications between users and producers would also be useful in the encouragement of the diffusion of innovations. But, while one may rely upon the ordinary forces of the marketplace to bring about a rapid diffusion of an existing innovation with good profit prospects, one can hardly rely completely upon such forces for the initial generation of such innovations.

References

Abernathy, W. J., and P. L. Townsend, 1975. Technology, Productivity, and Process Change, *Technological Forecasting and Social Change,* December.

Baker, N. R., et al., 1967. The Effects of Perceived Needs and Means on the Generation of Ideas for Industrial Research and Development Projects, N. R. Baker, J. Siegman, and A. H. Rubenstein, *IEEE Transactions on Engineering Management,* December.

Baker, N. R., et al., 1971. The Relationship between Certain Characteristics of Industrial Research Projects and Their Subsequent Disposition, *IEEE Transactions on Engineering Management,* November.

Battelle Research Institute, 1973. *Interaction of Science and Technology in the Innovative Process: Some Case Studies,* Battelle Research Institute, processed.

Carter, C. F., and B. R. Williams, 1957. *Industry and Technical Progress: Factors Governing the Speed of Application of Science to Industry* (London: Oxford University Press).

Carter, C. F., and B. R. Williams, 1959. *Investment in Innovation* (London: Oxford University Press).

David, P. A., 1975. *Technical Choice, Innovation and Economic Growth* (Cambridge: Cambridge University Press).

Federal Trade Commission, 1977. *Staff Report on the Semi-conductor Industry: A Survey of Structure, Conduct, and Performance,* Federal Trade Commission, Bureau of Economics.

Fellner, W., 1962. Does the Market Direct the Relative Factor-Saving Effects of Technological Progress? in *The Rate and Direction of Inventive Activity,* Universities-National Bureau Committee for Economic Research (Princeton: Princeton University Press).

Gibbons, M., and P. J. Gummett, 1977. Recent Changes in Government Administration of Research and Development: A New Context for Innovation?, presented to the International Symposium on Industrial Innovation, Strathclyde University, September.

Gibbons, M., and R. Johnston, 1974. The Roles of Science in Technological Innovation, *Research Policy,* November.

Gilpin, R., 1975. *Technology, Economic Growth and International Competitiveness,* study prepared for the Subcommittee on Economic Growth of the Congressional Joint Economic Committee (Washington, D.C.: U.S. Government Printing Office).

Golding, A. M., 1971. *The Semiconductor Industry in Britain and the United States: A Case Study in Innovation, Growth and the Diffusion of Technology,* D. Phil. thesis, University of Sussex.

Greenberg, D. S., 1966. "Hindsight": DoD Study Examines Return on Investment in Research, *Science,* 18 November.

Habakkuk, H. J., 1962. *American and British Technology in the Nineteenth Century* (Cambridge: Cambridge University Press).

Hicks, J. R., 1932. *The Theory of Wages* (London: Macmillan).

Hirshleifer, J., 1971. The Private and Social Value of Information and the Reward to Inventive Activity, *American Economic Review,* September 1971.

Illinois Institute of Technology, 1968. *Technology in Retrospect and Critical Events in Science,* processed.

Kennedy, C., 1964. Induced Bias in Innovation and the Theory of Distribution, *Economic Journal,* September.

Langrish, J., et al., 1972. *Wealth from Knowledge: A Study of Innovation in Industry,* J. Langrish, M. Gibbons, W. G. Evans, and F. R. Jevons (New York: Halsted/John Wiley).

Langrish, J., 1974. The Changing Relationship between Science and Technology, *Nature,* 23 August.

Langrish, J., 1977. Technological Determinism, processed.

Myers, S., and D. G. Marquis, 1969. *Successful Industrial Innovation* (Washington, D.C.: National Science Foundation).

Nelson, R., 1959. The Simple Economics of Basic Scientific Research, *Journal of Political Economy,* June.

Nelson, R. R., 1972. Issues in the Study of Industrial Organization in a Regime of Rapid Technical Change, in V. Fuchs, ed., *A Roundtable on Policy Issues and Research Opportunities in Industrial Organization* (New York: National Bureau of Economic Research).

Office of the Director of Defense Research and Engineering, 1969. *Project HINDSIGHT: Final Report* (Washington, D.C.: processed).

Pavitt, K./OECD, 1971. *The Conditions for Success in Technological Innovation* (Paris: OECD).

Pavitt, K., and W. Walker, 1976. Government Policies towards Industrial Innovation: A Review, *Research Policy.*

Peck, M. J., and F. M. Scherer, 1962. *The Weapons Acquisition Process* (Cambridge: Harvard University Press).

Rosenberg, N., 1969. The Direction of Technological Change: Inducement Mechanisms and Focusing Devices, *Economic Development and Cultural Change;* reprinted as Ch. 6 in Rosenberg (1976).

Rosenberg, N., 1974. Science, Invention and Economic Growth, *Economic Journal,* March; reprinted as Ch. 15 in Rosenberg (1976).

Rosenberg, N., 1976. *Perspectives on Technology* (Cambridge University Press).

Rothbarth, E., 1946. Causes of the Superior Efficiency of U.S.A. Industry as Compared with British Industry, *Economic Journal.*

Rothwell, R. et al., 1974. SAPPHO Updated: Project SAPPHO Phase II, R. Rothwell, C. Freeman, A. Horsley, V. T. P. Jervis, A. B. Robertson, and J. Townsend, *Research Policy,* November.

Rothwell, R., et al., 1976. Methodological Aspects of Innovation Research: Lessons from a Comparison of Project SAPPHO and FIP, R. Rothwell, J. Townsend, M. Teubal, and P. T. Speller, Maurice Falk Institute for Economic Research in Israel, *Discussion Paper 765,* processed.

Salter, W. E. G., 1966. *Productivity and Technical Change,* 2nd ed. (Cambridge: Cambridge University Press).

Samuelson, P. A., 1965. A Theory of Induced Innovation Along Kennedy–Weiszacker Lines, *Review of Economics and Statistics,* November.

Scherer, F. M., 1965. Firm Size and Patented Inventions, *American Economic Review,* December.

Schmookler, J., 1966. *Invention and Economic Growth* (Cambridge: Harvard University Press).

Sherwin, S. W., and R. S. Isenson, 1967. Project Hindsight, *Science,* 23, June.

Simon, H., 1966. Theories of Decision Making in Economics and Behavioral Science, *Surveys of Economic Theory,* vol. 3 (New York: St. Martin's).

Temin, P., 1966. Labor Scarcity and the Problem of American Industrial Efficiency in the 1850's, *Journal of Economic History,* September.

Teubal, M., et al., 1976. Performance in the Israeli Electronics Industry: A Case Study of Biomedical Instrumentation, M. Teubal, N. Arnon, and M. Trachtenberg, *Research Policy*.

Teubal, M., 1977. On User Needs and Need Determination: Aspects of the Theory of Technological Innovation, Maurice Falk Institute for Economic Research in Israel, *Discussion Paper* 774; processed.

Tilton, J. F., 1971. *International Diffusion of Technology: The Case of Semi-Conductors* (Washington, D.C.: Brookings).

Utterback, J. M., 1974. Innovation in Industry and the Diffusion of Technology, *Science*, 15 February.

von Hippel, E., 1976. The Dominant Role of Users in the Scientific Instrument Innovation Process, *Research Policy*.

von Hippel, E., 1977. The Dominant Role of the User in Semiconductor and Electronic Subassembly Process Innovation, *IEEE Transactions on Engineering Management*, May.

PART IV

Technology transfer and leadership: the international context

11 The international transfer of technology: implications for the industrialized countries

I

The first point to be made is that the transfer of technology from one place to another is not just a recent phenomenon but has existed throughout recorded history. Abundant archaeological evidence convincingly demonstrates that such transfer was an important aspect of prehistoric societies as well. Francis Bacon observed almost 400 years ago that three great mechanical inventions – printing, gunpowder, and the compass – had "changed the whole face and state of things throughout the world; the first in literature, the second in warfare, the third in navigation."[1] What Bacon did not observe was that none of these inventions, which so changed the course of human history, had originated in Europe, although it was from that continent that their worldwide effects began to spread. Rather, these inventions represented successful instances of technology transfer – possibly, in all three cases, from China.

It may be seriously argued that, historically, European receptivity to new technologies, and the capacity to assimilate them whatever their origin, has been as important as inventiveness itself. For inventions, as opposed to other goods, must be produced only once. And it is a conspicuous feature of their history that Europeans engaged in aggressive borrowing of inventions and techniques that had originated in other cultures.[2]

Thus, in the discussion that follows, we will not focus upon an activity that is unique to the second half of the twentieth century, although

This paper was prepared as a background paper for *North/South Technology Transfer: The Adjustments Ahead* (Paris: Organisation for Economic Cooperation and Development, 1981). Reprinted by permission.

[1] Francis Bacon, *The New Organon* (The Bobbs-Merrill Co., Indianapolis, 1960), p. 118.

[2] This point has been emphasized by A. R. Hall, among others. See his "Epilogue: The Rise of the West," in Charles Singer et al., eds., *A History of Technology*, vol. III (Oxford University Press, New York and London, 1957).

we will be interested in the special circumstances of recent technology transfers. Because such transfers have had a long history, some initial discussion of that history may be useful in gaining a better perspective on the present.

II

Although the international transfer of technology has been going on for a long time, the scale and the impact of such activities have vastly accelerated in the past 150 years or so. The industrial revolution, beginning in Great Britain in the last third of the eighteenth century, had at its center a rapidly expanding armamentarium of new technologies involving new power sources, new techniques of metallurgy and machine making, and new modes of transportation. These new technologies, when successfully organized and administered, brought immense improvements in productivity that transformed the lives of all participants.

One essential aspect of this revolution in technology must be understood because it is likely to be highly relevant to the prospects for the successful transfer of technology. The separate innovations – in metallurgy, power generation, and transportation – were, in significant ways, interrelated and mutually reinforcing. Often, one innovation could not be extensively exploited in the absence of others or the introduction of one innovation made others more effective. Metallurgical improvements, for example, were absolutely indispensable to the construction of more efficient steam engines. The steam engine, in turn, was utilized for introducing a hot blast of air into the blast furnace. The hot blast, by improving the efficiency of the combustion process, lowered fuel requirements and thereby reduced the price of iron. Thus, cheaper metal meant cheaper power, and cheaper power was translated into even cheaper metal. Similarly, the availability of cheap iron was essential to the construction of railroads. Once in place, however, the railroads reduced the considerable cost of transporting coal and iron ore to a single location. In this fashion, railroads reduced the cost of making iron. But cheaper iron, in turn, meant cheaper rails; this involved a further lowering of transportation costs, which again decreased the cost of producing iron. Thus, part of the secret of the vast productivity improvements associated with the new industrial technology was that the separate innovations were often interrelated and mutually reinforcing.

Although this transformation, which we call the industrial revolution, began in Britain, there was never any doubt that such new technologies would spread and be adopted elsewhere when the circumstances and surrounding conditions permitted (or were created). Indeed, British attempts in the nineteenth century to prolong their monopoly over these new technologies by legislative prohibition failed abysmally.

British technologies provided the basis for industrial development, first to Western Europe, then to the United States and, later, to other selected countries where conditions were favorable. British skilled labor, entrepreneurship and, sometimes, capital played a critical role in the early stages in bringing the new textile, transport, power, and engineering technologies to Western Europe.[3] The recipients of British technology were, therefore, in a distinctly favorable position. They could industrialize through the mere transfer of already existing technologies, without having to reinvent them. This ability to industrialize through borrowing rather than independent invention is the basic advantage of being a latecomer. However, the importance of this advantage should not be exaggerated. For one thing, the coin has two sides. Economic coexistence with advanced industrial societies entails a continual threat: Sophisticated, dynamic technology in the possession of such societies will generate innovations with very deleterious consequences to the less developed countries. The twentieth century provides numerous examples of the substitution of new products for old ones upon which some less developed countries had been heavily dependent – synthetic fibers for cotton and wool, plastics for leather, some nonferrous metals and other natural products, synthetic for natural rubber, synthetic detergents for vegetable oils in the manufacture of soap, and so on. An economy with no command over advanced technologies may be highly vulnerable to sudden changes in demand generated by these technologies abroad, and may have only limited opportunities for adjusting.[4] Moreover, the transfer of technology has never been easy. Typically, high levels of skill and technical competence are needed in the recipient country. It is hardly a coincidence that, in the nineteenth and early twentieth centuries, the countries that were most successful in borrowing foreign technologies were those that had well-educated populations (see Table 1). Furthermore, technologies are more than bits of disembodied hardware. They function within societies where their usefulness

[3] For an excellent account, see W. O. Henderson, *Britain and Industrial Europe, 1750–1870*, Liverpool University Press, Liverpool, 1954). Later, the French performed many similar functions in introducing the new technologies elsewhere in Europe. See Rondo Cameron, *France and the Economic Development of Europe* (Princeton University Press, Princeton, N.J., 1961).

[4] On the other hand, it should also be remembered that it was technical change and economic growth in the industrializing world that created the initial demand and profitable market opportunities for many primary products in the first place. In addition, synthetic products generate a demand for other primary products upon which they are based.

Table 1. *Estimated percentage of total population enrolled in school, selected countries in Europe, Northern America, and Oceania, 1830–1954*

	(1) 1830	(2) 1850	(3) 1878	(4) 1887	(5) 1928	(6) 1954
England and Wales		12		16	16	15
Scotland	9	—	15	16	17	17
Ireland		7		14	18	20
USA	15	18	19	22	24	22
Canada	6	—	21	20	—	20
Australia		—		14	16	17
Germany	17	16	17	18	17	13
Switzerland	13	—	15	18	13	13
Netherlands	12	13	16	14	19	20
Belgium		12		11	15	15
Denmark		14		12	16	16
Norway	14	14	14	13	17	14
Sweden		13		15	13	15
Finland	—	—	—	17	13	17
France	7	10	13	15	11	15
Austria	5	7	9	13	14	13
Hungary		7		12	16	14
Italy	3	—	7	11	11	13
Poland	—	—	—	—	14	13
Czechoslovakia	—	—	—	—	16	16
Spain	4	—	8	11	11	11
Portugal	—	1	—	5	6	11
Yugoslavia	—	—	—	3	10	13
Bulgaria	—	—	—	9	12	15
Rumania	—	—	—	2	12	13
Greece	—	5	—	6	12	15
Russia/USSR	—	2	2	3	12	15

Source: Richard A. Easterlin, "A Note on the Evidence of History," in C. Arnold Anderson and Mary Jean Bowman, eds., *Education and Economic Development* (Aldine Publishing Company, Chicago, 1965), p. 426. For detailed notes and qualifications concerning the data, see pp. 426–7.

is dependent upon managerial skills, upon organizational structures, and upon the operation of incentive systems. In addition, of course, a high payoff to the transfer of technology will depend upon the compatibility of its factor proportions requirements with those prevailing in the specific country or available nearby. It may also depend upon the quality of the country's natural resources – the phosphorus content of its iron ore, the richness of its copper ores, the sulfur content of its coal.

Thus, the successful transfer of technology is not a matter of transporting a piece of hardware from one geographic location to another. It often involves much more subtle issues of selection and discrimination, and a capacity to adapt and modify before the technology can function effectively in the new socioeconomic environment. Even apparently minor differences in resource quality, for example, may necessitate major alterations in the technology – as has often been the case for technologies involving chemical processes. These caveats, based upon widespread nineteenth-century experiences, are intended to suggest that the successful transfer of technology depends greatly upon the specific domestic circumstances in the recipient country.

But that is, obviously, only one part of the story. If we take the long view, it is apparent, first of all, that the transfer of industrial technology has been going on for as long as that technology has existed. We can observe its transfer, during the nineteenth century, to France, Germany, and the Low Countries, spreading gradually south and east. By the end of the century, this technology had spread to Russia and Japan. Both of these latecomers were to undergo extremely rapid industrialization in the twentieth century, although along entirely different institutional paths. Elsewhere in the world, the transfer of technology was closely associated with European migration and settlement patterns – first to the United States (which had a substantial manufacturing capacity as early as 1850[5]) and then to Canada, Australia, and New Zealand, brought by settlers from northwestern Europe, and later to parts of Latin America that were experiencing heavy immigration from southern and eastern Europe. With the exception of Japan – a remarkable exception – the technology transfers occurred earliest and most successfully in those societies that had close contacts with the industrializing nations of Europe.

Our central concern here, however, is to examine the impact of technology transfer upon the country of origin. First, therefore, we will make some observations about the long-term experience of the British economy. Because Britain was the first to develop and master industrial technology, and became its first major source of export, this nation

[5] Indeed, by the time of the Crystal Palace Exhibition in London in 1851, the British were so impressed with the distinctively American manufacturing technology that they dubbed it "the American system of manufactures." In the next few years, a substantial reverse flow of machine technology, from the United States to Britain, began. See Nathan Rosenberg, ed., *The American System of Manufactures* (Edinburgh University Press, Edinburgh, 1969). By the 1870s and 1880s, American multinational firms, following Singer Sewing Machine's leadership in opening a factory in Glasgow in the late 1860s, were undertaking foreign investments in significant numbers. See Mira Wilkins, *The Emergence of Multinational Enterprise* (Harvard University Press, Cambridge, Mass., 1970).

provides the most extensive body of evidence for examination. The British experience is also interesting because, unlike the experience of some later industrializers such as the United States, the success of the British economy has been linked to heavy foreign trade and, in particular, to the success of its export performance. It is also true, of course, that the British economy possessed certain features that should not be casually attributed to other OECD countries. Moreover, at a later stage, we will examine the relevance of the past to the economic circumstances of the present and future. Nevertheless, a brief consideration of the British experience will help to identify some of the variables of concern to us. We justify this assertion by observing that industrialization and the growth of per capita income have followed broadly similar trends wherever they have occurred.

III

The expansion of British industry in the nineteenth century involved drastic sectoral reallocations of the labor force. Whereas "Agriculture, Forestry, and Fishing" accounted for 35.9 percent of the working population in 1801, that figure had declined to 8.7 percent by 1901. "Manufacture, Mining, and Industry," which already accounted for 29.7 percent of the labor force in 1801, grew to 46.3 percent in 1901. This growth was not uniform over time but tended to occur in spurts. The most important growth took place in the second decade of the century, with substantial increases also in the 1840s and 1890s. During the nineteenth century, and linked to the expansion of industry, "Trade and Transport" grew from 11.2 percent of the occupied population in 1801 to 21.4 percent in 1901.[6]

These changes in the structure of the British economy were intimately connected with an enormous growth in the volume of world trade and with Britain's emergence as a highly specialized producer within the expanded market network. Even if it was an exaggeration to call Britain "The workshop of the world," by midcentury this nation was the main supplier of the quintessential nineteenth-century industrial commodities: iron, coal, railroad equipment, steam engines, and cotton textiles.[7]

[6] Phyllis Deane and W. A. Cole, *British Economic Growth, 1688–1959* (Cambridge University Press, Cambridge, 1969), p. 142. By 1951, the agricultural sector had declined to only 5 percent of the labor force, the manufacturing sector had grown to 49.1 percent, and trade and transport, although experiencing significant compositional changes, had increased only very slightly to 21.8 percent.

[7] Britain's role as a major raw materials exporter is usually neglected. In fact, Britain was, for many years, the world's largest coal exporter. At their peak in 1913, exports accounted for one-third of the United Kingdom's output of 287 million tons. See Deane and Cole, *British Economic Growth*, p. 216.

The last sentence suggests a key factor in the growth of world trade: a sequence of transportation innovations that opened up continental hinterlands, reduced the cost of the transoceanic shipping, and made possible the preservation of perishable food products during extensive voyages over land and sea. These innovations included the rapid growth of the railroads after 1830, the expanding role of the iron steamship after 1850,[8] and the introduction of refrigeration on both freight cars and steamships beginning in the 1870s (by the 1880s, shipments of frozen meats were reaching London from Australia). The reduced cost of ocean transport was not caused entirely by technological change in ship design. Various factors, such as decreased idle time in port, resulted in increased utilization of ship capacity, and the flow of immigrants out of Europe in midcentury brought a great improvement in the load factor. Moreover, industrialization and the rise of European populations and per capita income created a growing demand for food products and certain raw materials. The opening up of new agricultural lands and the decline of international transport costs also had a substantial impact upon European agriculture. With European wheat and corn no longer competitive at free-trade prices, an increased population shift from agriculture to manufacturing and the services resulted from the depression in European agriculture. In some areas, such as Prussia, these economic changes brought important political developments in their wake.[9]

It is important to understand, however, that the industrial revolution in the core countries was not, as is widely believed, heavily dependent upon imported raw materials from the periphery. As W. A. Lewis has pointed out:

[8] "From about 1860 . . . keen commercial competition existed between wooden vessels and iron vessels, between steam and sail, and this resulted in the driving down of freight rates in the course of the 1870's to unprecedentedly low levels; then, after about 1878, an entirely new range of cost economies was opened up by the use of metal-built steamships using steel hulls, steel boilers, twin screws, and compound engines. Freight rates for sailing ships on Atlantic passages fell from very remunerative to barely remunerative levels between 1874 and 1877–8, and then fell a further 40 percent . . . this eliminated the competition of sailing ships. In the East Indian trade they competed for rather longer. But by 1903 freight rates in general were down to about 20 percent of the 1877–8 level, and the actual costs of ocean shipment had fallen by an even greater percentage due to reductions in the cost of insurance." A. J. Youngson, "The Opening Up of New Territories," Chapter III in H. J. Habakkuk and M. Postan, eds., *The Cambridge Economic History of Europe,* vol. VI, *The Industrial Revolutions and After,* Part I (Cambridge University Press, Cambridge, 1965), p. 171.

[9] Alexander Gerschenkron, *Bread and Democracy in Germany* (University of California Press, Berkeley, 1943).

> The raw materials of the industrial revolution were coal, iron ore, cotton and wool; the foodstuff was wheat. All these the core produced for itself in abundance, with the United States and Europe complementing each other. Their chief deficiency was in wool, through which Argentina and Australia received their stimuli. Apart from this the core's principal imports in 1850 were palm oil, furs, hides and skins, a little timber, tea, coffee and other commodities in small quantities. It is hardly an exaggeration to say that the industrial revolution in the core did not depend on the periphery.[10]

Reduced transportation costs affect trade and development in another significant way. As the costs of moving materials from one location to another decline, the productive activity undertaken in any particular area is less intimately linked to the resources found there. It becomes increasingly possible to base industrial activity upon raw materials transported over a considerable distance, especially where there is convenient access to water transport. Thus, today, Japan's excellent steel industry is based upon both imported coal and imported iron ore.

IV

Some statistics may be cited to indicate the growing importance of exports to the British economy. Domestic exports as a percentage of national income ranged from 9 to 11 percent in the first half of the nineteenth century. This figure rose to a peak in the early 1870s, when domestic exports constituted about 23 percent of the national income. British exports began to encounter serious foreign competition in the 1880s, and thereafter fluctuated from 15 to 21 percent of the national income until the outbreak of the First World War. With growing problems in the interwar period, merchandise exports, which stood at about 17 percent of the national income in the late 1920s, fell to a mere 10 percent a decade later, and then recovered to the 20 percent level in the 1950s.[11] Thus, the size of the foreign trade sector is a crucial influence on the British economy. As Deane and Cole have concluded:

> It is not too much to say that the foreign-trade sector was setting the pace for British economic growth. It is thus not surprising that when the volume of British international trade ceased to grow, and even

[10] W. A. Lewis, *Growth and Fluctuations, 1870–1913* (Allen & Unwin, London, 1978), p. 30. Lewis, of course, recognizes that further technological change and the development of new industries changed this situation by the end of the nineteenth century.

[11] Deane and Cole, *British Economic Growth*, p. 29.

Table 2. *The pattern of imports into the United Kingdom, 1840–1950: current values as a percentage of total imports*

	Food, drink, and tobacco	Raw materials and semimanufactured goods	Manufactured and miscellaneous goods
1840	39.7	56.6	3.7
1860	38.1	56.5	5.5
1880	44.1	38.6	17.3
1900	42.1	32.9	25.0
1910	38.0	38.5	23.5
1930	45.5	24.0	29.4
1950	39.5	38.2	22.3

Source: Phyllis Deane and W. A. Cole, *British Economic Growth 1688–1959* (Cambridge University Press, Cambridge, 1962), p. 33.

declined, in the interwar period the pace of growth in real incomes fell to less than half of its peak nineteenth century rates.[12]

The growing dependence of the British economy upon export markets in the nineteenth century was intimately linked to the new industrial technologies. After 1850, with a new policy of free trade, Britain became increasingly dependent upon imported food supplies. The free trade policy also opened up the British economy to the manufactured products of industrial economies abroad, first from Western Europe and later from the United States. Manufactured goods constituted less than 6 percent of British imports in 1860 but accounted for 25 percent by 1900 (see Table 2). The economy achieved its well-known modern characteristics of a highly specialized manufacturing economy, heavily dependent upon overseas markets for its manufactured products in order to finance an unusually high import ratio – which reached a peak of about 36 percent of the national income in the early 1880s.[13]

The aggregate figures used so far, however, conceal some critical aspects of the use of industrial technologies in Great Britain. The essential point is that Britain came to rely upon a very narrow range of industries in its export activities – industries that had been the first to

[12] Ibid., pp. 311–12. Schlöte's estimates of the rate of growth of British exports are as follows: For the period 1780–1800 the annual average percentage rate of growth of exports was 6.1; for 1800–1825, 1.2; for 1825–1840, 4.0; for 1840–1860, 5.3; for 1860–1870, 4.4; for 1870–1890, 2.1; for 1890–1900, 0.7. W. Schlöte, *British Overseas Trade from 1700 to the 1930's* (Blackwell Scientific Publications, Oxford, 1952), p. 42.

[13] Ibid., p. 311. Imports fluctuated, after reaching this peak level, within the narrow range of 29 to 32 percent of national income until the outbreak of the First World War. The ratio fell to around 20 percent in the 1930s (ibid.).

make the dramatic change to centralized factory production brought about by the new industrial technologies. As later experience was to demonstrate, these were also the industries whose technologies could be most readily transferred to countries in the early stages of industrial development. Indeed, the major problem confronting cotton textiles, Britain's largest export industry, was not the competition of exporters from other countries, but the widespread process of import substitution, which reduced the proportion of cotton goods entering into worldwide trade.

The most serious loss of export markets for Lancashire textiles occurred in the lower-quality product lines. This is precisely what one would expect in a "product-cycle" view of international trade and comparative advantage. In this analytical framework, the least sophisticated mass-produced good is the first to have its production transferred abroad. Viewed in this light, the decline of the British textile industry reflects an inability to adapt, an incapacity to move aggressively into higher-quality cloth manufacture. This inability to adapt may also have reflected the relatively weak technical infrastructure of the British textile industry.[14]

Britain's domination of world trade in manufactured goods in the nineteenth century, and the rapid overall growth of its economy, which was so closely linked with the rapid expansion of these exports, had a narrow industrial base. Although, at the outbreak of the First World War, Britain was still the leading importer and exporter in the world, its relative position had been declining for some time. In large measure, this decline was inevitable. Indeed, it was an extraordinary aberration that this small island in the North Sea should *ever* have accounted for one-half of the world's total output of coal, over one-half of its pig iron (as it did in 1870), and more than three-quarters of the iron and steel products in international markets (even though British coal and iron ore deposits were particularly well suited to the events of the industrial revolution). Once this industrial technology began to be diffused abroad, as it inevitably was, the decline in Britain's share of world manufacturing production, as well as its share of trade in manufactured goods, was unavoidable (see Table 3). In the forty years before the First World War, the growth rate of manufacturing in the United States and Germany was more than twice that of the United Kingdom. In both countries also, it should be noted, the demand for basic industrial prod-

[14] See R. E. Tyson, "The Cotton Industry," in D. H. Aldcroft, ed., *The Development of British Industry and Foreign Competition, 1875–1914* (Allen & Unwin, London, 1968); Lars Sandberg, *Lancashire in Decline* (Ohio State University Press, Columbus, 1974).

Table 3. *Percentage distribution of the world's manufacturing production*

Period	United States	Germany	United Kingdom	France	Russia	Italy	Canada
1870	23.3	13.2	31.8	10.3	3.7	2.4	1.0
1881–5	28.6	13.9	26.6	8.6	3.4	2.4	1.3
1896–1900	30.1	16.6	19.5	7.1	5.0	2.7	1.4
1906–10	35.3	15.9	14.7	6.4	5.0	3.1	2.0
1913	35.8	15.7	14.0	6.4	5.5	2.7	2.3
1913[a]	35.8	14.3	14.1	7.0	4.4	2.7	2.3
1926–9	42.2	11.6	9.4	6.6	4.3	3.3	2.4
1936–8	32.2	10.7	9.2	4.5	18.5	2.7	2.0

Period	Belgium	Sweden	Finland	Japan	India	Other countries	World
1870	2.9	0.4	–		11.0		100.0
1881–5	2.5	0.6	0.1		12.0		100.0
1896–1900	2.2	1.1	0.3	0.6	1.1	12.3	100.0
1906–10	2.0	1.1	0.3	1.0	1.2	12.0	100.0
1913	2.1	1.0	0.3	1.2	1.1	11.9	100.0
1913[a]	2.1	1.0	0.3	1.2	1.1	13.7	100.0
1926–9	1.9	1.0	0.4	2.5	1.2	13.2	100.0
1936–8	1.3	1.3	0.5	3.5	1.4	12.2	100.0

[a]The second line for 1913 represents the distribution according to the frontiers established after the 1914–18 war.
Source: Folke Hilgerdt, *Industrialization and Foreign Trade* (League of Nations, Geneva, 1945), p. 13.

ucts such as iron and steel was growing far more rapidly than in Britain, and a "Chinese wall" of tariffs confronted British iron and steel products. The very size of the American market and its rapid rate of growth, in particular, meant that export markets were far less significant to American than to British industry.

V

One particularly important aspect of the British experience is this: Britain's dominance of world trade in the mid-nineteenth century, and to a declining extent thereafter, was rooted in the technologies that had created the original industrial revolution. Overwhelmingly, these technologies consisted of the output of the textile industry – especially cotton goods – and, to a much lesser degree, iron and steel manufactures.

Table 4. *The changing pattern of British commodity exports, 1830–1950: principal exports as a percentage of total domestic exports of the United Kingdom*

	1830	1850	1870	1890	1910	1930	1950
Cotton yarn and manufactures	50.8	39.6	35.8	28.2	24.4	15.3	7.3
Woollen yarn and manufactures	12.7	14.1	13.4	9.8	8.7	6.5	6.5
Linen yarn and manufactures	5.4	6.8	4.8	2.5	—	—	0.9
Silk[a]	1.4	1.5	0.7	1.0	0.5	0.3	2.3
Apparel	2.0	1.3	1.1	1.9	2.9	3.5	1.6
Iron and steel manufactures[b]	10.2	12.3	14.2	14.5	11.4	10.3	9.5
Machinery	0.5	0.8	1.5	3.0	6.8	8.2	14.3
Coal, coke, etc.	0.5	1.8	2.8	7.2	8.7	8.6	5.3
Earthenware and glass	2.2	1.7	1.3	1.3	1.0	2.1	2.5
Vehicles[c]	—	—	1.1	3.5	3.8	9.0	18.6
Chemicals	—	0.5	0.6	2.2	4.3	3.8	5.0
Electrical apparatus	—	—	—	—	—	2.1	3.9

[a]Including artificial silk.
[b]Including hardwares and cutlery.
[c]Carriages, wagons, ships, cars, cycles, aircraft.
Source: Phyllis Deane and W. A. Cole, *British Economic Growth, 1688–1959* (Cambridge University Press, Cambridge, 1969), p. 31.

In 1830, textiles alone accounted for almost three-quarters of all British exports, and the cotton sector by itself accounted for more than one-half (see Table 4). Textiles, plus iron and steel manufactures (which were far less important), in the same year accounted for almost 85 percent of the total exports. Cotton textiles, the largest sector and the first to be mechanized, began its long decline even before midcentury. The industry's dependence upon export markets is apparent in the fact that, on the eve of the First World War, three-quarters of its output was still being sold in foreign markets.

Whereas British exports were heavily based upon the industries that had pioneered the industrial revolution, the British economy was much slower in exploiting the new industries that began to emerge in the late nineteenth and early twentieth centuries, and that are sometimes referred to collectively as the "second industrial revolution."[15] To a much greater degree than the older industries, these new industries involved

[15] In some of the older industrial sectors there was also a conspicuous reluctance to adopt new techniques – such as in the iron and steel industry, where Britain became the world's largest importer by 1913, boot and shoe machinery, and certain portions of engineering and machine making. Britain's lagging export performance thus reflected a slow pace not only in the new and growing industries but often in the older, declining ones as well.

some degree of scientific knowledge. They included electricity and its multitudinous applications in engineering and machinery, organic chemistry and its ramifications, and the internal combustion engine. Britain's early industrial leadership did not translate readily into the new product classes that have played an increasingly important role in the twentieth century. This was true even where, as in the synthetic dye industry, the initial breakthroughs had occurred first in Britain.[16] It was only in the decades after the First World War that the British economy seriously confronted the structural transformations involved in shifting to the new industries that had been pioneered elsewhere – especially by the Germans, Americans and French.[17] In the meantime, as W. A. Lewis has pointed out, "organic chemicals became a German industry; the motor car was pioneered in France and mass-produced in the United States; Britain lagged in the use of electricity, depended on foreign firms established there, and took only a small share of the export market. The telephone, the typewriter, the cash register and the diesel engine were all exploited by others."[18]

A particularly interesting aspect of these "new" industries is the way in which their technologies were exported to Britain. Whereas British foreign investment during this pre-1914 era was overwhelmingly portfolio investment, its automobile, chemical, and electrical industries contained important components of direct foreign investments, from Ford and Westinghouse in the United States and Mond and Siemens in Europe. The use of direct rather than portfolio investment by these firms foreshadowed the pattern of international investment and international technology transfers that would come to dominate world trade in the later twentieth century.

[16] Perkins achieved the synthesis of aniline purple dye (mauve) while working at the Royal College of Chemistry in London in 1854. See John J. Beer, *The Emergence of the German Dye Industry* (University of Illinois Press, Urbana, 1959).
[17] This characterization is, unavoidably, broad and sweeping. There was great variation in the performance of different industries, and some distinguished British successes, as in bicycles and textile machinery. For a more detailed industry-by-industry examination of British export performance before the First World War, see D. Aldcroft, ed., *The Development of British Industry and Foreign Competition, 1870–1914* (University of Toronto Press, Toronto, 1968). The chapters on the engineering industry, electrical products, and chemicals are particularly valuable in conveying the complexity and diversity of experience even within individual industries. For a useful discussion of technical change in Britain in the interwar years, see R. S. Sayers, "The Springs of Technical Progress in Britain, 1919–1939," *Economic Journal*, June 1950.
[18] Lewis, *Growth and Fluctuations*, p. 130.

VI

Britain's slowness to use major new innovations highlights a central concern of this paper. Historically, much of the impact of new technologies has been associated with the rise of new industries producing new products. Indeed, economic growth itself has been largely determined by the capacity to use new technologies, whether developed at home or abroad. That association is not an adventitous event; as Simon Kuznets has richly documented, strong economic growth is a reflection of a continuous shift in product and industry mix. All rapidly growing industries eventually slow down as the cost-reducing impact of technological innovation diminishes. Continuous rapid growth thus requires the development of new products. The reason is that, given the typically low long-term income and price elasticity of demand for old final consumer goods, further cost-reducing innovations in those industries will have a small impact. As Kuznets has expressed it:

> [A] sustained high rate of growth depends upon a continuous emergence of new inventions and innovations, providing the bases for new industries whose high rates of growth compensate for the inevitable slowing down in the rate of invention and innovation, and upon the economic effects of both, which retard the rates of growth of the older industries. A high rate of over-all growth in an economy is thus necessarily accompanied by considerable shifting in relative importance among industries, as the old decline and the new increase in relative weight in the nation's output.[19]

Thus, Britain's deteriorating economic performance in the twentieth century is not to be measured, as it is rather naively at times, by its declining relative importance in world trade in manufactured goods. That decline was unavoidable, given the combination of rapid industrialization abroad and the obvious supply constraints upon a small country. Rather, Britain's deteriorating performance may be observed in its inability to generate or exploit new technologies with anything like the success achieved in the coal, iron, and steam technologies of the original industrial revolution. A vast literature has appeared on the reasons for this decline, including the question of whether, and in what sense, it constituted a failure.[20] For the present, we only wish to emphasize that, in the twentieth century, the British economy responded slowly and

[19] Simon Kuznets, *Six Lectures on Economic Growth* (The Free Press, Glencoe, Ill., 1959), p. 33. For the original, detailed expression of this view, see Simon Kuznets, *Secular Movements in Production and Prices* (Houghton Mifflin Co., Boston, 1930).

[20] A useful entry point into this literature is provided by D. McCloskey, ed., *Essays on a Mature Economy: Britain after 1840* (London, 1971).

sluggishly to the structural transformations required by the decline of old industries and the expansion of new ones.[21] Industries such as textiles, coal, steel, and shipbuilding failed to achieve either sufficient reduction in size or technical modernization, where the latter was possible. (One might well argue that the preferential treatment accorded to British exports in its protected imperial markets postponed the need for these essential changes and thus harmed the British economy.[22]) The pressures to make such structural transformations became increasingly urgent in the twentieth century as wars, depression, and the growth of import substitution abroad drastically changed the export markets for many of the traditional older industries. Economic success for Britain, and for industrial Europe in general, turned upon the ability to respond to rapid changes in the structure of world trade.[23]

The viability of an economy – its capacity to respond to changing market circumstances – is significant for a whole range of issues. The ability to adapt to changing conditions helps to determine the success or failure of economic performance. A viable economy will find opportuni-

[21] As W. A. Lewis has stated of Britain's lackluster performance: "it is not necessary to be a pioneer in order to have a large export trade. It is sufficient to be a quick imitator. Britain would have done well enough if she merely imitated German and American innovations. Japan, Belgium, and Switzerland owe more of their success as exporters of manufactures to imitation than they do to innovation." W. A. Lewis, "International Competition in Manufactures," *American Economic Review Papers and Proceedings*, May 1957.

[22] Indeed, Kindleberger has stated of British export performance in the period 1875 to 1913: "Britain in this period had too high a rate of growth of exports, rather than too little, since exports of standard goods to the countries of the Empire enabled the economy to evade the exigencies of dynamic change, away from cotton textiles, iron and steel rails, galvanized iron sheets, and the like, to production for export or for the home market of the products of the new industries." Charles Kindleberger, "Germany's Overtaking of Britain, 1860–1914," *Weltwirtschaftliches Archiv*, 1975, p. 491.

[23] This theme, of the inability to achieve structural transformation, is dealt with in detail by Ingvar Svennilson, *Growth and Stagnation in the European Economy* (United Nations Economic Commission for Europe, Geneva, 1954). Svennilson argues, moreover, that the capacity to institute such changes is a declining function of industrial maturity. "There is no doubt that in the oldest and most advanced industrial countries, transformation of the economy in accordance with new trends in technology and demand meets with strong resistance from the accumulated stock of capital, from the traditional special skill of labour – in fact, from the whole organization of society. The resistance to the penetration of electricity in old economies built up round the coal-fields and based on the steam engine is a good illustration of this tendency . . . For the less-advanced countries, the opportunities for industrial growth were large both in manpower and in techniques which could easily be imitated, while the resistance to transformation offered by an old industrial structure was either weaker or non existent." Ibid., p. 206. For further discussion of some of these issues, see Ed Ames and Nathan Rosenberg, "Changing Technological Leadership and Industrial Growth," *Economic Journal*, March 1963.

ties and growth stimuli in circumstances where an economy lacking the capacity to change will encounter only "obstacles" to growth. In this sense, it is important not to exaggerate the role of foreign trade as an independent stimulus. For, as Kindleberger has pointed out:

> one cannot really discuss the role of foreign trade in growth without indicating the underlying capacity of the economy to undertake new tasks in depth or to transform . . . It is not the foreign trade that leads to growth – any stimulus can do it if the capacity to transform is present or can be drawn out of dormancy. Without that capacity, a benign stimulus may lull the economy into senescence and a rude one may set it back. With it, both can lead to growth.[24]

VII

In discussing the impact of technology transfer upon the country of origin, certain distinctions are highly important. Perhaps the most basic question is whether these technologies occur in industries that compete directly with those of the initiating country, or whether the relationship between the technologies is complementary. Much of the technology that was transferred from Britain to Western Europe, for example, was in competitive industries, whereas most of the technology transferred abroad in the period 1870–1913, the heyday of British foreign investment, was in industries that complemented British industry. Let us consider the latter case first.

As Britain developed into a mature industrial nation in the second half of the nineteenth century, it profoundly influenced the world economy. One of the striking features of this impact was a very large outflow of foreign investment, much of it incorporating recent technological developments, between 1870 and 1913. In this period, such foreign investment amounted to about 4 percent of Britain's national income, although it was a good deal higher toward the end of the era – about 7 percent between 1905 and 1913. This huge outflow served, among

[24] Charles Kindleberger, *Economic Growth in France and Britain* (Harvard University Press, Cambridge, Mass., 1964), p. 287. See also Charles Kindleberger, "Group Behavior and International Trade," *Journal of Political Economy*, February 1951. In that article, Kindleberger argued that the conventional analytical tools of the economist had to be supplemented by a theory of group behavior in order to make adequate predictions concerning the response to economic stimuli originating in international trade. Although the article is primarily concerned with the agricultural sector, Kindleberger offers the more general suggestion "that the flexibility of a society in devising institutions to accomplish its purposes under changing conditions is a function of its social cohesion, which, in turn, depends upon its internal social mobility, system of communications, and set of values." Ibid., p. 45.

other things, as a powerful vehicle for the transfer of technology. What can be said about this experience for present purposes?

First of all, the outflow of British capital took the form of portfolio rather than direct investment. As a result, the recipients of this capital retained considerable freedom in using the funds and in determining who would supply their industrial technologies. Although the products of the new industrial technology often turned out to be British, that was not necessarily the case. In this respect, British portfolio investment was structured very differently from the more recent direct foreign investment activity of the multinational firm, which retains control over the transfer of industrial technology to the recipient country. In this particular and perhaps crucial respect, therefore, British foreign investment before the First World War offered a degree of freedom (when the investments were made by individuals rather than firms) that is typically absent today.

In addition, most British capital flowed to a limited number of regions. These were not the densely settled portions of the world where Leninist theory predicted that advanced capitalist societies would move in their relentless search for cheap labor. Quite the contrary. Only a small fraction of this investment went to such areas. The great bulk of it went to the thinly populated regions of recent European settlement. Typically, these were areas of abundant and unexploited natural resources with great potential for producing and exporting primary products to Britain – the United States, Canada, Argentina, Australia, New Zealand, South Africa. Moreover, only a small fraction of this foreign investment was received by firms which were mainly oriented toward the production of commodities for foreign markets. Most of it was used to expand the flow of primary products to the rapidly growing markets of Great Britain. (The swift decline of British agriculture after the move to free trade in the 1840s had further increased the demand for imported primary products.) British investment and technology was embodied primarily in infrastructural facilities – railroads, dock and harbor installations, and other public utilities, as well as in firms engaged in the extraction of raw materials for export. It is estimated that, by 1913, slightly over 40 percent of British foreign investment was in railways alone, and approximately 15 percent was in mines and raw materials. An additional 30 percent was in government loans, a large fraction of which was also in railways, mines, and raw materials.[25]

These investments and technology transfers, then, went into produc-

[25] For further details on British foreign investment, see A. C. Cairncross, *Home and Foreign Investment, 1870–1913* (Cambridge University Press, Cambridge, 1953).

tive activities which stood in a strongly complementary relationship to the needs of the British economy. In the last couple of decades of the nineteenth century, real income in Britain benefited considerably from the resulting improvement in the British terms of trade, as the prices of food and raw materials sharply declined.[26]

Even when exported technologies are directed toward industries that compete with rather than complement those of the home country, the ultimate outcome may not be easy to specify. The adjustments required by the expansion of competitive industries abroad are very different in a world economy where demand is growing rapidly – as it was during much of the nineteenth century or in the twenty-five years after the Second World War – as compared with one where markets are stagnating or growing only slowly – as was the case in the interwar years. Moreover, from many points of view, critical social and economic issues may turn upon the capacity to make the necessary adjustments. The ability to do this may be partly a matter of scale; an economy may be able to make certain kinds of structural adjustments more readily when these adjustments are small rather than large. Britain has experienced extremely painful adjustment problems in the twentieth century for two reasons. In 1900, the country had not only a very large export trade but one in which well over half of its exports were concentrated in declining industrial sectors. (It is also true, however, that Britain experienced a decreasing share of certain commodity classes that were becoming increasingly important in world trade. This was true, for example, of iron, steel, and engineering.)

Recent history suggests that even this is much too parochial a view. It is, of course, true that the transfer of industrial technology to developing countries and import substitution lead to a declining importance of manufactured goods in the import basket of these countries. Table 5 clearly shows the declining share of finished manufactures in U.S. imports in the second half of the nineteenth century, and Table 6 shows the changing importance of specific categories of imported goods as a percentage of domestic consumption. But that is far from the entire story. This transformation was associated with a substantial increase in real income per capita (as well as population), as a result of which the absolute amount of imported manufactured goods increased substantially. The view of world trade as consisting primarily of poor countries exporting food and raw materials in exchange for the manufactured goods of industrial countries is both incomplete and distorted when it is presented

[26] For details, see Charles Kindleberger, *The Terms of Trade* (John Wiley, New York, 1956).

Table 5. *Share of finished manufactures in U.S. imports*

Period	Percent
1964–1968	45.1
1954–1963	30.7
1944–1953	18.6
1939–1948	17.7
1929–1938	22.2
1919–1928	18.8
1909–1918	17.7
1904–1913	21.4
1899–1908	22.4
1889–1898	24.8
1879–1888	28.3
1869–1878	34.6
1859–1868	43.5
1850–1858	47.2
1850	54.6
1840	44.9
1830	57.1
1820–1821	56.4
1770	n.a.

Source: L. Davis et al., *American Economic Growth* (Harper & Row, New York, 1972), p. 568.

as the entire story.[27] The fact is that industrial nations are each other's best customers, and the increasing specialization of manufacturing in the twentieth century has created a broad economic basis for trade among them. Indeed, it is important to observe that, in 1900, when the British were already deeply concerned over the increasing commercial competition of Germany and the United States as those countries rapidly industrialized, the two major purchasers of British manufactured goods were – Germany and the United States. As Folke Hilgerdt pointed out long ago, "countries which increased their manufacturing more than the world average as a rule also increased their imports of manufactured goods more than the world average"[28] (see Table 7).

[27] See Albert Hirschman, *National Power and the Structure of Foreign Trade*, (University of California Press, Berkeley, 1945). Hirschman showed that, for the period 1925–37, about half of the manufactured goods entering world trade were exchanged for other manufactured goods. The other half were traded for foodstuffs and raw materials.

[28] Folke Hilgerdt, *Industrialization and Foreign Trade* (League of Nations, Geneva, 1945), pp. 93–4.

Table 6. *Imports as a percentage of U.S. consumption*

	1869	1909	1947
Manufacturing, total	14.0	5.9	2.2
Foods	19.8	9.5	3.8
Beverages	15.2	5.8	1.8
Tobacco products	5.3	3.0	.1
Textile products	20.8	8.6	1.7
Leather products	4.0	2.1	1.0
Rubber products	10.0	1.0	.1
Paper products	32.8	5.8	8.3
Printing and publishing	2.8	1.4	.1
Chemicals	26.8	11.8	2.5
Petroleum and coal products	0	.6	1.4
Stone, glass, and clay roducts	11.7	5.5	1.1
Forest products	3.6	3.6	3.5
Iron and steel products	12.0	1.4	.2
Nonferrous metal products	20.1	9.2	5.2
Machinery	.9	.8	.3
Transportation equipment	0	.8	.2
Miscellaneous	17.2	5.7	3.1
Agriculture	5.8	8.3	6.0
Fisheries	1.1	4.8	15.3
Mining	2.1	7.3	6.0
Total	10.9	6.8	3.2

Source: L. Davis et al., *American Economic Growth* (Harper & Row, New York, 1972), p. 572.

VIII

Thus, industrialization has drastically increased the volume of world trade, although there have been major counteracting forces, such as the Great Depression and widespread recourse to restrictive commercial policies. The reasons for the resulting growth of trade are not far to seek. The ability of poor countries to import manufactured goods has always been severely constrained by their poverty and by their capacity – or, rather, incapacity – to earn foreign exchange. Although technology transfer and industrialization abroad have characteristically reduced the market for specific categories of manufactured products – such as textiles and other light consumer goods – this import substitution process has been more than offset by the growing demand for other kinds of manufactured products resulting from income growth and the altering needs of industrializing (and, therefore, expanding) economies. Many of the problems confronting the advanced industrial countries

Table 7. *Movement of manufacturing and trade in manufactured articles, up to 1926–9*

Country	1926–9 as percentage of 1891–5			1926–9 as pecentage of 1901–5		
		Trade in manufactured articles (quantum)			Trade in manufactured articles (quantum)	
	Manufacturing	Imports	Exports	Manufacturing	Imports	Exports
Japan	1,932	628	1,660	659	240	588
Finland	583	473	280	325	273	175
Canada	521	—	—	318	284	—
New Zealand	—	—	—	296	226	—
Italy	394	189	583	254	135	246
United States	436	230	803	250	202	338
India	—	—	—	213	170	—
France	260	127	177	205	98	138
Belgium	285	—	—	204	120	144
Sweden	405	480	426	192	201	337
Germany	279	185	203	163	149	133
United Kingdom and Ireland	143	195	144	120	126	122
World	326	225	230	207	165	165

Note: The countries are arranged in the order of the increase in their manufacturing activity between 1901–5 and 1926–9.
Source: Folke Hilgerdt, *Industrialization and Foreign Trade* (League of Nations, Geneva, 1945), p. 93.

during this process have resulted not from a declining demand for manufactured products but from their inability to adapt to changing demands. Britain's slowness in moving out of its old specializations in railway vehicles and ships as world demand declined in the interwar years, and its slow exploitation of the rapidly growing market for road vehicles, are cases in point.

This change in the composition of world trade is one of the most distinctive features of the twentieth century; it is central to an understanding of the impact of technology transfer in the present day and, doubtless, in the years ahead. Maizels's data on the commodity pattern of world trade (Table 8), although highly aggregated, show clearly the massive decline in textile products, which accounted for over 40 percent of world trade in manufactures in 1899 and only 11 percent in 1959. This decline reflects the widespread practice of import substitution as

Table 8. *Commodity pattern of world trade in manufactures, 1899–1959 (percent)*

	Excluding Netherlands			Including Netherlands					
	1899	1913	1929	1929	1937	1950	1955	1957	1959
Metals	11.5	13.7	12.1	11.9	15.3	12.9	15.0	15.5	13.5
Machinery	8.0	10.4	13.9	14.5	16.0	20.7	22.3	24.1	24.8
Transport equipment	3.8	5.4	9.9	9.8	10.5	14.2	15.3	16.1	16.5
Passenger road vehicles	0.3	1.7	3.7	3.6	3.8	4.1	4.8	4.7	6.0
Other transport equipment	3.5	3.7	6.2	6.2	6.7	10.1	10.4	11.4	10.5
Other metal goods	7.0	6.5	5.9	5.9	6.5	4.9	4.8	4.4	4.2
Chemicals	8.3	9.1	8.4	8.5	10.6	10.5	11.4	11.0	12.0
Textiles and clothing	40.6	34.1	28.7	28.7	21.5	19.9	13.4	11.8	11.1
Yarns	9.0	6.4	5.0	5.0	4.1	3.7	2.5	2.2	2.0
Fabrics	23.2	21.2	17.5	17.5	12.9	12.2	7.7	6.7	6.1
Made-up goods	8.4	6.5	6.3	6.2	4.6	3.9	3.2	2.9	3.0
Other manufactures	20.8	20.7	21.1	21.0	19.5	17.0	17.9	17.1	17.9
Total	100.0	100.0	100.0	100.0	100.0	100.0	100.0	100.0	100.0
Total in $ billion	3.11	6.50	11.90	12.20	9.27	20.77	34.37	43.48	45.89

Source: A. Maizels, *Industrial Growth and World Trade* (Cambridge University Press, Cambridge, 1965), p. 164.

Western textile technology was diffused abroad, although the relatively low income elasticity of demand for textile products in any case precluded any growth in world trade in textiles commensurate with that of other manufactured goods. The impact of this import substitution process was most severe in Britain, the largest exporter of textile products during the interwar years. Britain was also adversely affected by the growing competition from other textile suppliers for the reduced volume of world trade.[29] Japan, in particular, became a major competitor in the interwar years. The U.S. share of the textile market rose sharply after the Second World War as the Americans successfully exploited their advantage in the technology of synthetic fibers.[30] Offsetting the decline of trade in textiles that has been associated with the industrialization process is a major increase in the share of world trade consisting of capital equipment, chemicals, and the more complex and sophisticated range of consumer goods. Thus, whereas machinery and transport equipment constituted less than 12 percent of world trade in manufactures at the turn of the century, it accounted for over 25 percent of the total in 1937 and over 40 percent by 1959. Machinery alone, which accounted for 8 percent in 1899, accounted for one-quarter of the total in 1959. The growing world trade in capital goods reflects not only the critical importance of capital equipment to the development process but the additional vital fact that import substitution has, in general, been relatively unimportant in capital goods. It is only at a more mature stage of industrial development that the large-scale production of capital goods is undertaken.[31]

Several of the trends described in the previous paragraph are cap-

[29] See A. Maizels, *Industrial Growth and World Trade* (Cambridge University Press, Cambridge, 1965), chap. XIII.

[30] Unlike cotton textiles, which have relatively modest capital requirements – an important consideration in the timing of the import-substitution process – synthetic fiber production involves large capital investments per unit of output.

[31] Indeed, there has been a high degree of concentration, internationally, in the production of capital goods. "The supply of capital equipment on the world market has remained considerably more concentrated in the hands of the 'big three' industrial countries – the United States, Britain and Germany – than has that of other manufactures. In 1913, for example, some four-fifths of all capital goods exports came from the big three, and the proportion was the same in both 1929 and 1937; by 1950 it had fallen to three-quarters, and there was a further small fall, to 70 percent, by 1959. By contrast, the proportion of other manufactures exported by the big three was only two-thirds in 1913 and fell to one-half by 1959. Only large industrial countries export a full range of capital equipment on a substantial scale. Exports from the smaller industrial countries tend to be more specialized; for example Swiss machine tools of advanced design, or Swedish electrical apparatus." Maizels, *Industrial Growth*, p. 276.

tured, in microcosm, in the changing pattern of British trade relations with India.

> Before the first World War, the Indian import market was very largely a British preserve. Britain supplied 85 percent of India's imports of manufactures in 1899 and 80 percent in 1913. The greater part of the trade at that time was in cottons and other textile manufactures. The growth of factory production of cotton textiles in India in the 1920's was already adversely affecting the level of imports, and Britain was the main loser. There was a further sharp contraction in the 1930's, and by 1937 Britain's textile exports to India were one one-seventh of their 1913 volume. Since Independence, India has emerged as a major exporter of cotton goods, her imports being negligible. On the other hand, Britain has made good some part of her loss of the textile market by increasing her share of India's imports of machinery, transport equipment and chemicals; this group accounted for almost three-quarters of total 'imports' of manufactures into the Indian sub-Continent in 1959 compared with only one-quarter in 1929.[32]

Although the relative importance of consumer goods has declined in world trade, this has not been the case for consumer durables. Such goods generally have a high income elasticity of demand;[33] therefore, rising incomes in developing countries have meant a disproportionately large increase in the demand for such goods. Moreover, many of these goods cannot be easily produced until a relatively mature stage of industrialization is attained, so that rising demand has usually been associated with an increase in world trade. After reviewing the production statistics for the major categories of consumer durables, Maizels concludes:

> First, production of durable consumer goods has made considerable strides in many overseas markets since the early 1950's; in particular, there has been rapid development in the economically more advanced countries, such as Australia and New Zealand, and also more recently in some Latin American countries, such as Argentina and Mexico. Secondly, these countries have concentrated on the less complex products, such as bicycles and refrigerators, and also on those which are suitable for large-scale assembly operations such as motor cars and sewing machines.[34]

[32] Ibid., pp. 195–8.

[33] There are exceptions. At sufficiently high income levels, bicycles appear to become "inferior" goods.

[34] Maizels, *Industrial Growth,* pp. 317–18. After the Second World War, British exports suffered particularly badly from import substitution in Australia, New Zealand, and South Africa.

The great growth in world trade in consumer durables, especially since the Second World War, has occurred primarily among the industrial nations themselves, and not between the industrial and the less developed worlds. Partly, of course, this is because many expensive durables can be purchased only by affluent consumers; partly, it is because of the reduction or elimination of trade restrictions in the industrial countries after the Second World War and the resulting increased specialization in these countries, involving greater attention to refinements in styling and other qualitative variations in product, permitting a higher ratio of trade to output.[35] As a result, the location of the markets for expensive durables, combined with the technical complexities, high skill requirements, and economies of scale often involved in their production, have confined the production of such commodities to a small, although expanding, group of advanced countries. This concentration has been further reinforced by the fact that these countries continue to be the locus of technological dynamism and new product innovation.

Nevertheless, the less developed world has steadily expanded its manufacturing sector. Even before the twentieth century, there were many such undertakings, often abortive, in countries that are still considered less developed – such as India, Mexico, and Brazil.[36] The process was accelerated during the two world wars, when the supply of industrial products was either seriously disrupted or cut off. Import substitution typically began with small-scale production activities requiring only modest amounts of capital and skilled labor. And, of course, the sine qua non has always been the existence of a sizable local market for the product. Textiles, again, has been the most likely early candidate, but

[35] It is also true that small countries tend to be more narrowly specialized than larger ones and are thus more dependent upon trade for expanding their range of goods.

[36] This is perhaps an appropriate place to voice objection to the simple dichotomies – or trichotomies – that dominate so much of the present discussion of industrialization and technology transfer. To place all of the nonindustrial countries of the world in a single category – whether it be "less developed," "Third World," or, perhaps least satisfactory of all, "developing" (because the central problem in many "developing" countries is precisely that they are *not* developing) – does not contribute to better understanding. There is a tremendous diversity of conditions that is only obscured by conceptual devices that place Brazil and Nepal, or India and Uganda, in the same category. At the very least, it must be recognized that, if our criterion is the size of the industrial sector, there is a continuum within the category of so-called less developed countries that is so broad that few propositions that are true of one end of that continuum will also be true of the other end. At the more advanced end of that continuum are many countries now successfully engaged in the process of industrializing – Brazil, Mexico, India, South Korea, Taiwan. Some of these countries, at least, have far more in common with the industrialized countries than with the least industrialized countries.

other light consumer goods have also been important; food processing based on local products and the fabrication of locally available raw materials have also been early contenders. At a later stage, when higher incomes and expanded markets have justified the introduction of more complex and expensive consumer durables, it has been common to enter new industries first by assembling components produced abroad – radios, refrigerators, washing machines, and automobiles – and then by gradually increasing the reliance upon locally produced components. This gradual moving up the scale of technical complexity by less developed countries has been going on for generations; there is no reason to expect the process to change.

IX

One central conclusion that may be drawn from this account is that the transfer of industrial technology to less developed countries is inevitable. Indeed, as we have seen, the process has already been going on for about a century and a half, and there is no compelling reason to believe that it will stop. Not only is industrialization elsewhere inevitable, but so is some alteration in the relative positions of the industrial countries. Britain's overwhelming industrial leadership in the 1830s had been largely eroded by the Germans and the Americans by the outbreak of the First World War, and Japan has emerged as one of the great industrial powers in the last quarter of a century. Thus, a historical perspective suggests that the central questions are not whether industrial technologies will be transferred, but rather when it will happen, where it will happen, which technologies will be transferred, how they will be modified in the process, and how rapidly this process will occur.

One of the most interesting aspects of the history of technology transfer is the difficulty of controlling or containing its spread. Great Britain was totally unsuccessful in its attempts, in the first half of the nineteenth century, to prolong its monopoly over the new technology. The British, however, have not been the only ones to express this view. The Germans and the Americans did the same when they later achieved technological leadership. Moreover, the spread of industrial technology has been uncontrolled in a quite different sense. The findings or outcomes of the research process are impossible to predict, and it has often happened that research created discoveries of great benefit in unexpected – and unintended – places. The introduction of the basic lining into the blast furnace by two English chemists, Gilchrist and Thomas, in the late 1870s was of immense benefit to Britain's continental competitors because it vastly expanded the range of ores that could be utilized in modern steel-making technologies. The Gilchrist–Thomas technique made possible the intensive exploitation of Europe's great phosphoric ore deposits and thus provided the technological basis for the great expansion of steel production in Germany and Belgium after 1880.

Perhaps the most distinctive single factor determining the success of technology transfer is the early emergence of an indigenous technological capacity. In the absence of such a capacity, foreign technologies have not usually flourished. Countries that have had successful experiences usually learned at an early stage that the importation of foreign technologies required some minimum level of technological skills – not only to modify and adapt the foreign technology to local needs, once it had been imported, but to provide the basis for an intelligent selection among the wide range of potential foreign suppliers. Intelligent *choice* among the alternative technologies available abroad presupposes considerable technical knowledge. Such knowledge, in turn, is difficult to acquire in the absence of any domestic experience or capacity.

The Japanese experience has been particularly instructive in this respect. The Japanese were notably successful in adapting Western technology to their very different factor proportions. Moreover, it may be highly significant that they did so in ways that involved almost no reliance upon either foreign enterprise or foreign direct investment. In general, the Japanese opposed any arrangements that reduced local control over the technology.

In a variety of ways, the Japanese adapted Western technology so as to reduce the capital-output ratio. In textiles, for example, they purchased older, secondhand machines – often machines that had been discarded as obsolete in Lancashire. Moreover, once installed, they operated the machinery at higher speeds and for longer hours than in England or America, and they lavished greater amounts of labor on servicing the machines and maintaining them in a decent state of repair. When the Japanese eventually built their own textile machines, they substituted wood for iron wherever possible – for example, in the beams. They introduced cheaper raw materials into production, as in the case of cotton spinning, and then added more labor to each spinning machine to handle the increased frequency of broken threads. In addition, Western techniques were introduced in highly selective ways at particular stages, and old-fashioned, capital-saving, cottage-industry techniques continued to be used in other stages – as in raw silk production and cotton weaving. In a variety of ways, Japanese industry continued to make important use of relatively labor-intensive, small-scale industry through subcontracting arrangements. Auxiliary activities such as materials handling and packaging were far more likely to continue to be carried out by hand labor.[37]

Similar kinds of "capital-stretching" adaptations could be enumerated from the industrial experiences of Korea and Taiwan after the Second World War.[38] But with respect to these highly successful adaptations of Western technology, we must emphasize that it is often a serious mistake to associate a fixed factor intensity with a given piece of industrial hardware. In many instances, the hardware itself *will* dictate a particular factor intensity, with little room for substitution – especially in productive processes where economies of scale are important, as in petroleum refining and other chemical processing industries.[39] But the recent experience of some of the most successful Asian economies indicates that there are many opportunities for substituting labor for capital when using Western industrial machinery. Furthermore, as Raymond Vernon has cogently argued, the factor intensity of production and skill requirements will generally vary considerably over the life cycle of any given product.[40] Although abundant unskilled or semiskilled labor may offer only limited opportunities in the early stages of new-product development (when skill requirements are typically high), they may offer much greater cost advantages when the product matures and the production technology has stabilized.

Thus, the transfer of technology must not be conceived of as a once-and-for-all affair. It is not something that happens at a single point in time. It is, rather, an ongoing activity. Any perspective that ignores this fact is likely to distort the essential issues in technology transfer. The successful transplantation of a technology involves the domestic capacity to alter, modify, and adapt in a thousand different ways – often in ways that are subtle and evident only to a person with considerable

[37] See G. Ranis, "Factor Proportions in Japanese Economic Development," *American Economic Review*, September 1957, and the same author's recent article, "Industrial Sector Labor Absorption," *Economic Development and Cultural Change*, April 1973. For some evidence on the experience with the use of secondhand machinery in Mexico and Puerto Rico, see W. Paul Strassmann, *Technological Change and Economic Development* (Cornell University Press, Ithaca, N.Y., 1968, chap. 6. For a discussion of the rather extensive Brazilian experience with secondhand capital equipment, see N. Leff, *The Brazilian Capital Goods Industry 1929–64*) Harvard University Press, Cambridge, Mass., 1968). The notion that there is something "inferior" about the use of secondhand equipment needs to be dispelled. The Greeks have developed one of the world's most successful shipping industries while relying very heavily upon secondhand ships.

[38] Ranis, "Industrial Sector Labor Absorption," pp. 402–8.

[39] For a valuable empirical discussion of economies of scale, see C. F. Pratten, *Economies of Scale in Manufacturing Industry* (Cambridge University Press, Cambridge, 1971).

[40] Raymond Vernon, "International Investment and International Trade in the Product Cycle," *Quarterly Journal of Economics*, May 1966.

technical expertise. An economy that lacks the domestic capacity to do these things is most unlikely to make successful use of innovations developed far away and in response to a very different set of circumstances. Conversely, an economy that possesses or can acquire this capacity is in a position to draw upon more advanced technologies abroad in ways that can yield spectacular results. This capacity has been critical to the remarkable performance of Japanese industry in the twentieth century, especially in the past twenty-five years. The Japanese have elevated to a fine art what they call *improvement engineering*. Although this kind of technical skill was not responsible for any major, original inventions, it has enabled them to draw upon a large inventory of foreign technologies and reshape them to their own requirements with a high degree of sophistication. And, after all, what increasingly matters in a highly integrated world economy is not inventive ability but the capacity to exploit new technological opportunities, whatever their country of origin. As Rosovsky has pointed out:

> Improvement engineering reduces the real cost of imported technology by making it more productive. At the margin, it will permit the adoption of some techniques that otherwise would not have been profitable to operate. If, by means of these efforts a technique is made more economical, it could also result in a greater domestic and foreign market, leading to otherwise unexploitable economies of scale. The very act of improvement engineering can also raise the quality of certain categories of workers. It is largely an activity of "carefully taking apart and putting together a little better"; it is concrete and directly related to production, especially when compared to basic research. In contrast to the pursuit of core innovations, this type of activity is less risky and much cheaper: in effect, one is working in already proven directions. What is the message for underdeveloped countries? Simply this: Western technology need not be treated as a given; simple and small improvements suited to local conditions are frequently possible.[41]

[41] Henry Rosovsky, "What Are the 'Lessons' of Japanese Economic History?," in A. J. Youngson, ed., *Economic Development in the Long Run* (St. Martin's Press, New York, 1972), pp. 248–9. On some larger themes, Rosovsky states: "In considering the sweep of Japanese economic history, I am struck by the notion that government–business relations were, from the local point of view, well arranged. Japan retained some advantages of capitalism (i.e. efficient producers), while reaping certain benefits of socialism (i.e. considerable public control of the economic effort and direction)." Ibid., p. 249. Rosovsky also points out that, although Japan was indeed economically backward by any of the ordinary indicators at the time of the Meiji Restoration, it "possessed some rather unusual non-economic attributes: a comparatively literate population, fertility control within marriage, an effective central and local government, an efficient system of roads and communications, large cities, excellent art, architecture, literature, etc." Ibid., p. 232.

Table 9. *R&D spending in five industrial countries*

Year	France	West Germany	Japan	U.K.	U.S.
Ratio of all R&D expenditures to GNP					
1961	1.38	NA	1.39	2.39	2.74
1962	1.46	1.25	1.47	NA	2.73
1963	1.55	1.41	1.44	NA	2.87
1964	1.81	1.57	1.48	2.30	2.97
1965	2.01	1.73	1.54	NA	2.91
1966	2.03	1.81	1.48	2.32	2.90
1967	2.13	1.97	1.53	2.33	2.91
1968	2.08	1.97	1.61	2.29	2.83
1969	1.94	2.05	1.65	2.23	2.74
1970	1.91	2.18	1.79	NA	2.64
1971	1.90	2.38	1.84	NA	2.50
1972	1.86	2.33	1.85	2.06	2.43
1973	1.77	2.32	1.89	NA	2.34
1974	1.81	2.26	1.95	NA	2.32
1975	1.82	2.39	1.94	2.05	2.30
1976	1.78	2.28	1.94	NA	2.27
1977	1.79	2.26	NA	NA	2.27
1978	NA	2.28	NA	NA	2.25
Ratio of civilian R&D expenditures to GNP					
1961	0.97	NA	1.37	1.48	1.10
1962	1.03	1.14	1.46	NA	0.97
1963	1.10	1.26	1.43	NA	0.81
1964	1.34	1.38	1.47	1.46	1.02
1965	1.37	1.53	1.53	NA	1.00
1966	1.40	1.62	1.47	1.58	1.10
1967	1.50	1.70	1.51	1.68	1.23
1968	1.54	1.72	1.60	1.70	1.37
1969	1.49	1.81	1.64	1.69	1.47
1970	1.47	1.96	NA	NA	1.52
1971	1.37	2.16	NA	NA	1.41
1972	1.39	2.13	NA	1.49	1.44
1973	1.30	2.01	NA	NA	1.47
1974	1.34	2.27	1.91	NA	1.46
1975	1.41	2.20	NA	1.50	1.42
1976	1.42	2.09	NA	NA	1.39

Source: Science Indicators 1978 (National Science Board, Washington, D.C., 1979), pp. 140, 144.

Another important point must be made about the Japanese experience. The usual emphasis on Japan as an importer of foreign technology must not be allowed to osbscure the fact that the Japanese commit a large volume of resources to the search for productivity-raising new technologies. In fact, the Japanese economy in recent years has been highly research-intensive. Although Japan has consistently ranked behind the United States and the United Kingdom in R&D expenditures as a percentage of the gross national product, its position becomes much more favorable if we exclude the large commitment in those countries to purely military R&D (see Table 9). Moreover, private industry in Japan pays for a larger share of total R&D than does private industry in other industrial countries. Thus, Japan industry has made a heavy commitment of resources to the kinds of activities upon which the exploitation of sophisticated technology is dependent. It has devoted great attention to activities that have expanded its absorptive capacity.[42] The Japanese have continually stressed the growth of their own technological capabilities. Direct foreign investment has been virtually excluded, and advanced technologies have been acquired by relying heavily upon licensing agreements together with a large civilian R&D effort. Thus, a major ingredient of Japan's remarkable success story seems to have been a government strategy for introducing foreign technologies in ways that emphasized their local linkages and the emergence of an indigenous technological capacity.

X

What factors will shape the impact of this developing process, with its concomitant increasing competition confronting the exports of the industrial countries, or displacing their production for their own domestic markets? To answer this question, we would have to consider not only the forces that have operated in the past, as we have done in this chapter, but also the ways in which novel elements, or elements more important in the present than in the past, are likely to function. We can do no more here than call attention to some of these elements.

[42] "The level and pattern of research and development within Japan are closely related to the import of technology from abroad. Firms must maintain some research capacity in order to know what technology is available for purchase or copy, and they must generally modify and adapt foreign technology in putting it to use – a 1963 survey of Japanese manufacturers showed that on average one-third of the respondents' expenditures on research and development went for this purpose . . . The moderate level, wide diffusion, and applied character of Japan's research effort are consistent with a facility for securing new knowledge from abroad." Richard Caves and Masu Uekusa, *Industrial Organization in Japan* (The Brookings Institution, Washington, D.C., 1976), p. 126.

The first observation – abundantly confirmed by recent experience, especially since the Second World War – is that the successful transfer of industrial technology need not mean a reduction in the volume of foreign trade or an absolute decline in the export opportunities confronting any individual country. Trade is not a zero-sum game in which new participants increase their industrial exports by depriving established industrial economies of an equivalent amount. It does, however, involve a readiness to accept changes in the composition of output, with all of the structural adjustments and other painful reallocations that may be entailed in changing trading patterns. Trade also means a willingness to accept new patterns of specialization as the industrializing countries establish new competencies and displace more advanced industrial economies from traditional niches based upon earlier but now eroding comparative advantages. These adjustments, however, are coming into increasing conflict with other domestic goals, such as high employment, strong expectations about the continued growth in real wages, and increasing sensitivity to the distributional aspects of adjustment processes when left to the marketplace. Successful policies dealing with displaced workers and in speeding their responses to changes in demand have proven exceedingly difficult to formulate and implement – and, of course, the difficulties themselves have been a main justification of the growth in protectionist sentiments.

In the future, one of the key elements in the impact upon the industrial countries when technology is transferred to the less developed countries will be the capacity of the industrial countries to continue to generate new technologies, especially new products – and the *rate* at which these can be generated. There are now powerful forces at work, many themselves the result of technological innovation – improved techniques of communication and transportation – that are speeding up the diffusion of new technologies from the center to the periphery. These centrifugal tendencies are powerfully assisted by the multinational firm, which, through its large-scale foreign investments and licensing activities, has become the most powerful institution for the spread of new technology in the post-World War II years.[43] Given the possibility for more rapid diffusion of technologies, the capacity to generate new tech-

[43] The preoccupation with the role of the multinational firm in "north-south" economic relations often distracts attention from the fact that most foreign investment of multinational firms has gone to *other* advanced industrial economies. This activity has been an integral part of the new patterns of industrial specialization that have come to characterize the most advanced economies in the past couple of decades. Thus, if we consider foreign direct investment of American enterprises in manufacturing subsidiaries in 1969, fully 73 percent of the total went to Europe and Canada, 15 percent to Latin America, and 12 percent went to all other areas. See Raymond Vernon, *Sovereignty at Bay* (Basic Books, New York, 1971), p. 65.

nologies will play an even greater role in the economic destinies of the industrial countries. This is because the time available to exploit their lead with respect to any given technology is bound to decline.[44]

There are, to be sure, offsetting forces at work. An increasing number of industrial technologies are becoming science-based – a situation strikingly different from the nineteenth century. As this happens, it may become increasingly difficult to borrow or imitate without a reasonably high-level domestic science capability – as is the case, to varying degrees, in chemicals, aircraft, telecommunications, and electronics. If this turns out to be so, it may be important, in slowing down the progression of the less industrial countries to increasingly complex, research-intensive production. Even in such cases, however, lead times are likely to decrease. The life cycle of many products demonstrates that, as even high-technology products mature and stabilize, new possibilities emerge that make their production more compatible with the relative factor prices and skills available in less advanced economies. In some cases, novel divisions of labor have emerged, leading to increasing reliance upon "offshore" contributions, at least for certain stages in the productive process, or for a growing number of components of increasingly complex final products. Elsewhere, there seem to be many opportunities for modifications in final product design or specification that will make high-technology products far more compatible with the capabilities of less advanced economies. Eventually, of course, increasing elements of science will be transferred abroad – as has already occurred in Japan – so that economic advantages grounded in a greater science capability will also prove to be ephemeral. Such advantages may be further eroded, in any case, by institutional as well as technological innovations that will provide substitutes to compensate for a limited domestic science capability.

[44] Multinational firms have shown a strong preference for direct foreign investment over other options, such as licensing, in the exploitation of their technological leads. This accounts for the rapid growth in overseas subsidiaries in the past quarter of a century. There are various reasons for this preference, including the numerous difficulties involved in dealing with the purchase and sale of information in the marketplace. As Richard Caves has observed: "The alternative of licensing a foreign producer can match the profitability of direct investment only in certain cases, namely where the rent-yielding advantage of the parent firm lies in some one-shot innovation of technique or product, such as a new method for making plate glass or the secret ingredient of a successful soft drink. Only in these cases can the information on which the parent's advantage rest be easily transferred intact to a foreign firm. In other cases, either the information cannot be transferred independent of the entrepreneurial manpower, or uncertainty about the value of the knowledge in the foreign market will preclude agreement on the terms of a licensing agreement that will capture the full expected value of the surplus available to the licensor." Richard Caves, "International Corporations: The Industrial Economics of Foreign Investment," *Economica*, February 1971, p. 7.

In this changed international environment, it is impossible to assert with any confidence what the final outcome is likely to be. There are more channels and mechanisms for the transfer of technology than existed a century ago, creating the possibility for a more rapid transfer. On the other hand, the technologies themselves are more complex and often enmeshed in a systemlike relationship with other hardware as well as software; this has important implications for the control of the technologies being transferred. In some industries, especially the processing activities, economies of scale are substantial. Moreover, there is an increasing dependence upon strong engineering and scientific skills.

These circumstances help to account for the growing prominence of the multinational firm in recent years as an institutional device for technology transfer. Indeed, what the multinational firm has often done has been to transfer entire technological packages – transfers that draw not only upon the technological capacities of such firms but upon their organizational and managerial abilities, their easy access to capital, and their extensive marketing skills. The rise of the multinational firm, with its advantages in exploiting managerial, financial, and communications resources, seems to have lubricated the wheels of international technology transfer.

Although certain forces thus seem to be facilitating the widespread diffusion of industrial technology, with its peculiar characteristics, they are also working to maintain centralized control and hierarchical relationships. This is true, in general, of the improvements in communication and the vastly expanded information-processing capacity of electronic technology, as well as the increased complexity and importance of scale economies in modern industrial technologies. All of these forces favor large producing units as well as expand the effectiveness of centralized management and the control of far-flung production units. In addition, as Raymond Vernon has pointed out:

> the cost of generating and launching major industrial innovations will continue to grow, relative to the other costs associated with the production and marketing of the products concerned. On the other hand, it also seems likely that the rate of adoption of innovations by industrial users and consumers will continue to accelerate, when compared with historic norms. So will the rate of appropriation and imitation on the part of the producers that pursue a follow-the-leader strategy. These tendencies, taken in combination, suggest that those multinational enterprises that base their business strategy on an innovational lead will have to plan even more than in the past for the speedy exploitation of any industrial advance over the largest possible market. This means that such enterprises will continue to place a high value on

quick and easy access to overseas markets and that they will constantly try to extend the geographical reach of their distribution network.[45] Obviously, the exact role of the multinational firm will be the outcome of political and ideological as well as economic factors.

Finally, there is the big question of the emergence of new goals, or at least shifting priorities, in the industrial countries. Over the past fifteen years or so, there has been an increasing commitment to improving the quality-of-life variables – the reduction of environmental pollution and a growing concern with occupational safety and health, and product safety. In numerous ways, direct and indirect, such new concerns have had the effect of raising the costs and reducing the incentives for technological innovation. There is evidence that these concerns have already had a strong effect upon productivity growth in the United States.[46] To maintain high rates of technological innovation – the main characteristic of leading industrial nations in the past – will require imaginative new directions, including institutional innovation and new incentive systems, if new priorities are to be successfully reconciled with continued technological leadership.

The tradeoffs between more rapid growth in production and environmental pollution will also eventually pose urgent policy questions for the less developed countries. So long as increased production of material goods continues to be stressed, one possibility is that the developing countries will increasingly specialize in the dirty, polluting industries that are being rejected by the advanced countries. Indeed, there is already evidence of such movement – as in the case of the asbestos industry. An even more desirable solution would be for the less developed countries to take advantage of the past experience of the industrial countries in order to deal with environmental problems in a more imaginative and successful fashion. Nevertheless, painful choices, in which a more rapid rate of growth in the output of material goods entails some environmental degradation, appear to be unavoidable.

[45] Vernon, *Sovereignty at Bay*, p. 252.

[46] See the suggestive calculations of Edward Denison, "Effects of Selected Changes in the Institutional and Human Environment upon Output per Unit of Input," *Survey of Current Business*, January 1978. In considering the impact of recent measures to deal with pollution, occupational health and safety, and crime, Denison finds that "By 1975 . . . output per unit of input in the nonresidential business sector of the economy was 1.8 percent smaller than it would have been if business had operated under 1967 conditions. Of this amount, 1.0 percent is ascribable to pollution abatement and 0.4 percent each to employee safety and health programs and to the increase in dishonesty and crime. The reductions had been small in 1968–70 but were rising rapidly in the 1970's. The increase in their size cut the annual change in output per unit of input from 1972 to 1973 by 0.2 percentage points, the change from 1973 to 1974 by 0.4 percentage points, and the change from 1974 to 1975 by 0.5 percentage points." Ibid., pp. 21–2.

12 U.S. technological leadership and foreign competition: *De te fabula narratur?*

I

I begin with a brief deck-clearing operation, one that I hope will be noncontroversial. The discussion of America's role in international markets for high-technology products was for a long time dominated by some highly unrealistic expectations. These expectations were formed in the years immediately following World War II, when a whole generation of Americans grew up surrounded by tangible evidence of America's across-the-board technological superiority. For twenty years or so following the Second World War, and for reasons closely connected with the uneven incidence of that war and its sequelae, American technological leadership was one of the prime facts of international life. The years from 1945 to the mid- or late 1960s were, without doubt, the age of American technological hegemony.

Precisely because of their comparative technological backwardness, however, the other Organization for Economic Cooperation and Development (OECD) member countries were able to combine their recovery, not only from wartime destruction but from a longer period of neglect stretching back to World War I, with a rapid rate of technical improvement. So long as there remained a substantial gap between technology levels in the United States and the other advanced industrial countries, there was a possibility of rapid technological change through the transfer and adoption of the more sophisticated and productive American technology.

Thus, throughout the 1950s and 1960s, and with varying degrees of effectiveness, the other OECD countries played a highly successful game of technological catchup. A combination of high rates of capital formation – on the average far higher than in the United States – plus the importation and exploitation of more advanced American technologies,

This paper was prepared for and presented to a National Academy of Sciences Panel on Advanced Technology Competition and the International Allies, December 1981. It is printed with the permission of the National Academy of Sciences. The views expressed do not necessarily reflect opinions of the Academy or the U.S. Government. Valuable comments by Moses Abramovitz and Claudia Frischtak are gratefully acknowledged.

brought a progressive narrowing of American technological leadership vis-à-vis Europe and Japan.

These developments also resulted in far higher rates of productivity growth abroad by comparison with the United States. These trends came to be regarded as alarming in America only in the 1970s, when our own rate of productivity growth, which had long trailed behind those abroad, fell precipitously after a perceptible but slower decline in the late 1960s.

At present, therefore, in many fields, America's earlier lonely eminence at numerous technological frontiers has passed. Today, other industrial nations have attained positions close to, or at, these same frontiers. In many ways, this situation should be cause for rejoicing: We are no longer living in the readily identifiable aftermath of the most destructive war in history. Although we are, perhaps understandably, preoccupied with the more competitive aspects of the situation, we need to be reminded that companionship at the technological frontier offers considerable benefits as well as costs. This brings me to one of my central themes: In the years immediately ahead, I expect to see a good deal of convergence in the economic environment of a sizeable number of countries. By contrast with the postwar years of American hegemony, we are likely to see several technological competitors functioning within increasingly similar economic environments and therefore responding to increasingly similar stimuli and problems.

It is very plausible to argue that the extent of U.S. technological leadership and higher income levels in the early postwar decades were so great that they limited our capacity to derive much benefit from the technological activities of other countries – at least in terms of the possibility of transferring useful technology, as opposed to the benefits that occurred through normal trade relations. From this point of view, the growing similarity of certain conditions in other industrial countries is increasing our capacity to derive such benefits. More rapidly rising labor costs in Japan and Western Europe and the increasing relative scarcity of certain raw materials in the United States contribute to this greater likelihood. Foreigners are increasingly responsive to innovative possibilities that will benefit us, such as labor-saving innovations. Moreover, resource-saving innovations, which have long been of greater concern to other countries, are becoming more relevant and applicable in the United States. This point is being forcefully underlined by the huge influx of more energy-efficient Japanese cars. More generally, as real incomes abroad rise more rapidly than in the United States, one may readily expect a greater number of attractive new products, designed mainly for affluent households, to enter the U.S. market from abroad.

As a result of these trends, it is already becoming difficult to determine whether some developments reflect further erosion of American technological leadership as opposed to other commercial considerations. Consider the recent rise in foreign patenting in the United States, which has often been cited as evidence of increasing inventive capability in other countries relative to the United States. It is doubtful that such data constitute good evidence of the changing relative pace of technological progress. It is more plausible to argue that the rising percentage of patents in the United States obtained by foreigners has been dominated by commercial judgments and considerations such as changes in the size of specific markets, in the composition of demand, in relative prices, and so on. They do not necessarily indicate changing technological capabilities. Some of these changes may reflect the impact of growing concern over environmental pollution and other hazards that have become embodied in government regulations. The rising sensitivity to pollution has considerably increased the proprietary value of high-performance automobile emission-control devices and has led to decisions to enter the American market with technologies that, in some cases, had been around for some time. The rise of energy prices may be drastically increasing the profitability of energy-conserving technologies in the American market. Because American energy prices have long been below those of the rest of the world, it is reasonable to believe that their dramatic rise in the 1970s increased considerably the profitability in the U.S. market of energy-saving devices or designs that were *already* in use in other countries. Such price changes would have the effect of making patents in the U.S. market more commercially attractive than they were previously. No change in relative inventiveness has necessarily taken place. Thus, increased patenting by foreigners in the U.S. market may indicate little about changes in inventiveness between the United States and the rest of the world and a great deal about the changing pattern of incentives generated by underlying economic forces. In *some* respects, of course, the final outcome may be the same, whether the intensifying competitive challenge to an American high-technology firm is the result of shifts in international technological capabilities or increasing similarity in the economic conditions of a growing number of industrial countries. From a policy point of view, however, correct diagnosis may make a great deal of difference.

The perspective I have laid out has an additional implication that deserves to be made explicit. The changed American international position is readily subject to a variety of alarmist interpretations. Indeed, there are aspects of our recent economic performance that I find particularly alarming – such as the extremely poor productivity growth of the American economy during the 1970s. Nevertheless, policy formulation is not well served by mindless extrapolations from the recent past. To the extent that I am correct in my analysis of technological catchup abroad and the growing similarity of conditions between the United States and those countries that have been successful in catching up, it would be unwarranted to expect the economic performance of those countries rapidly to surpass that of the United States. On the one hand, rapid technological change becomes much more difficult to sustain as the technological frontier is approached and reached. Catchup gives way to the more difficult and costly joint participation in the process of pushing out that frontier. Additionally, economies that arrive at affluence may be expected to share some of the features that slowed down the growth of the technological leader – including an increasing commitment to improving the quality of the environment, a growing sensitivity to safety and health and other "amenity" considerations, the eventual exhaustion of the labor pool from low-productivity agricultural occupations that could, by moving elsewhere, make substantial contributions to productivity growth, the growing size of the service sector, where productivity improvements seem more difficult to attain than in the commodity-producing sector, and so on. Thus, an alternative to the scenario of other industrial countries rapidly surpassing America in technological and economic leadership is that of a number of countries drifting, within a narrowing range, toward some asymptotic level of economic performance, a level first attained by America. As the gap between them narrows, the later arrivals come increasingly to assume some of the characteristics of the leader. As Marx, writing of the industrial revolution in England, warned his European readers in the preface to *Capital:* "The country that is more developed industrially only shows, to the less developed, the image of its own future." Or, as he warned them more succinctly in the previous paragraph: "De te fabula narratur!"[1]

[1] Karl Marx, *Capital* (The Modern Library, New York, n.d.), p. 13. Of course, England's disappointing economic performance in the century or so since the publication of *Capital* in 1867 does represent a conceivable scenario for the United States. In that case, the United States would be overtaken and surpassed by more vigorously growing economies, just as England was eventually surpassed by the United States, Germany, and Japan. The intriguing question here is whether there is a peculiarly English disease accounting for its lackluster performance or whether its experience is the reflection of more deeply rooted forces that will inevitably assault all countries at some sufficiently advanced stage of industrial maturity.

II

A central feature of high-technology industries that is likely to become increasingly significant is an apparently inexorable rise in the development costs of new products. In some measure, these rising costs are inevitable when complex, state-of-the-art products are being designed and when performance improvements are likely to bring great advantages. The move toward higher performance levels commonly involves higher development costs, because there is an identifiable tradeoff where better performance is available at higher cost (as, for example, in the decision to use more expensive materials). Additionally, however, the competitive process also drives firms to bring the new product into the market ahead of the competition. Greater speed in new product development – in the extreme case, the "crash program" approach – is inherently more costly than a slower, step-by-step, sequential process. This approach is most readily observable in advanced weapons development for the military and the space program, where costs are less important than performance improvement and where cost overruns are tolerated as an unavoidable fact of life. Although the beneficial spillovers from the military and space programs to the civilian sector are often cited, far less attention has been given to their possible deleterious effects in raising the costs of civilian R&D and in reducing the sensitivity of American engineers to cost considerations of a kind that are likely to be decisive in commercial markets.

The extreme impact of rising development costs is particularly apparent in the commercial aircraft industry, especially since the advent of the jet engine in the 1950s. For some years, commercial aircraft manufacturers were able to limit development costs by adopting new technologies only after they had been produced and operated for some years by the military. The Boeing 707 was a civilian version of the KC-135 military tanker, an aircraft that had been produced in large numbers for the military, and even the 747 benefited from the development experience Boeing derived from its unsuccessful bid in the C-5A competition. With the increased focus upon missiles, however, the military and commercial sectors have diverged. Commercial firms now confront costs of the order of magnitude of $1 billion in developing a new generation of widebodied jets and have less direct financial support from earlier military technologies. In 1981, McDonald-Douglas refused an offer by Delta Airlines to undertake the development of a new commercial aircraft, despite Delta's willingness to place an order of over $1.5 billion.

Thus, in the commercial aircraft industry, participating firms confront extremely high development costs in addition to the costs of production. This problem is compounded by the fact that the market for commercial aircraft is relatively small – in part, a testimony to the high productivity of commercial jets. Few commercial jets ever sell in excess of a few hundred units – and only two (the 727 and the DC-9) have ever sold in excess of 1,000.

Thus, the extreme commercial risks posed by high development costs are likely to dominate developments in this industry in the future. Subcontracting, at least partially as a risk-sharing device, is already an important aspect of the industry. Boeing had six major subcontractors for the 747, with whom it shared the development costs and risk, and the development of the aircraft began only after firm purchase commitments were in hand from Pan Am, TWA, Lufthansa, and BOAC. Boeing has subcontracting arrangements for its new-generation 767 and 757 with a number of foreign firms – Japanese, Canadian, Italian, British. One cannot help wondering to what extent such international subcontracting is also motivated by the perception that the arrangement will make it easier to enter those commercial markets.

High development costs and large financial risks also figure prominently in the increasingly European recourse to international consortia – as in the case of the successful Airbus and the ill-fated Concorde. Although there are presently only three commercial airframe manufacturers and two commercial jet engine manufacturers in the United States, the numbers are even smaller in Western Europe, where the industry is now largely nationalized.[2] In addition, it is important to observe that the Concorde, a brilliant engineering achievement but a commercial disaster (only sixteen were manufactured before production was discontinued), was made possible by immense subsidies from the French and British governments.

The huge development costs and the associated financial risk of commercial aircraft are paralleled by similar trends in other high-technology industries. There are obvious parallels in the field of military procurement – with the additional special problem of a single buyer. Development costs of nuclear power reactors, where safety and environmental considerations are especially important, have inexorably risen. More conventional power-generating equipment, although not plagued by the special problems of nuclear power, involves technological and other performance uncertainties that have resulted in very high development costs. The exploitation of new fossil-fuel energy sources, involving liq-

[2] Shortly after this was written (November 1981), Lockheed announced its intention to terminate the production of the wide-bodied L-1011. When that happens, there will be only two American commercial airframe manufacturers.

uefaction and gasification, is almost certain to encounter spectacular development costs, as is already clear at the pilot-plant stage. Telecommunications, involving extreme systemic complexity, compatibility, and interdependence problems, and the special features associated with network evolution, also encounters very high development costs; the cost of the no. 4 Electronic Switching System was about $400 million. Finally, although the electronics industry has some very different features from the industries just mentioned, the design and development of reliable, high-capacity memory chips has drastically raised the stakes for commercial survival. Almost daily reports in the financial press suggest that hundreds of millions of dollars in development costs are being incurred in the international competition for the 64K RAM market. The prospects for recovering these development costs have been jeopardized by widespread price cutting.

Thus, industries confronting a combination of dynamic, rapidly improving technology and high development costs may be expected to share many problems in varying degrees. Financial risks are becoming exceedingly great, in some cases requiring markets substantially larger than can be provided by a single moderately sized Western European country. For technological and other reasons (sometimes regulatory), long lead times are often involved; thus, the full recovery of an investment is deferred, at best, to the far distant future. Not only are technological uncertainties extreme, but the large financial commitments are frequently required during precisely that early stage when uncertainties are greatest. The very fact of rapid technological change raises the risk of investing in long-lived plant and equipment because further technological change is likely to render such capital soon obsolete. If, as is widely asserted, product life cycles are themselves becoming shorter, the agony of the risk-taking process is further intensified. What is certainly true is that in committing large amounts of resources to the development process, timing becomes even more crucial. There is abundant recent evidence that new, technologically complex products experience numerous difficulties in their early stages that may take years to iron out. The earliest Schumpeterian innovators frequently wind up in the bankruptcy courts. The strategy of a rapid imitator, or "fast second," benefiting from the mistakes of the pioneer has much to commend it, especially when rapid technological change is expected to continue. This was clearly the case with British pioneering of the commercial Comet I, well before American entry into the commercial jet age. As it happened, substantial improvements in engine performance in the next couple of years offered Boeing and Douglas decisive commercial advantages, in terms of greater capacity and speed, that were incorporated into their later entrants – the 707 and DC-8.

The development process eventually dovetails into investment in plant and equipment. Indeed, in a high-technology world of prototypes and pilot plants, it may be impossible to draw a sharp line of demarcation between the two. In other respects, however, high-technology industries are bringing the two stages even closer together. Increasingly in some of these industries, the performance of the final product and the efficiency of the productive process are extremely sensitive to the precise nature of the process technology involved in manufacturing. In electronics, yield and performance are closely linked to the maintenance of scrupulously high standards of cleanliness and other related aspects of quality control. Thus, innovation in electronics has created even greater intimacy between product and process innovation, which must now be considered together. The great increase in circuit-element density, leading to dramatic improvements in the capabilities of an integrated circuit chip, has been inseparable from the introduction of more complex processing equipment. This processing equipment, in turn, raised the fixed costs of a wafer fabrication plant from $2 million to $50 million during the 1970s. Such high-capacity plants, of course, need to operate at very large volumes.

Here again, high-technology innovation intensifies the need for expanding output and pushes producers beyond the limitations imposed by domestic markets. For a variety of reasons, including tariffs and other forms of protective legislation that place imported products at a competitive disadvantage, multinational firms may attempt to exploit foreign markets by establishing a local subsidiary. Such actions obviously accelerate the transfer of technology; indeed, multinational firms are now a main instrument for the more rapid international diffusion of advanced technologies.

In an increasingly interdependent world, it becomes more and more difficult to isolate the domestic and international contexts of decisions regarding innovation. Thus, it is not sufficient to argue that transferring a technology overseas has the unambiguous effect of narrowing the technology gap between the United States and the rest of the world. Obviously, in making a decision to develop a new technology in the first place, a high-technology firm will take into account the prospect of earning revenues from overseas as well as in domestic markets. How important these prospects are is a difficult matter of commercial judgment, but it seems apparent that American firms are more willing to undertake development expenditures at home if there is a possibility of eventually transferring the technology overseas. This factor, if at work, will strengthen the willingness to commit resources to the innovation process. In a world of shortening product life cycles and the more rapid diffusion of new technologies, the willingness to undertake domestic innovative activity may be strengthened by the prospect of overseas payoffs. And a speeding up of the international diffusion of new technologies seems to assure a shortening of the time period for an innovating firm to exploit its technological lead.

In an increasingly interdependent world with rapidly advancing technologies, we must also question the propriety of treating an "industry" as some ultimate or indivisible unit of analysis. Consider, for example, the dissolution of the boundary line that could once be drawn between the telecommunications and computer industries. The microchip revolution and the growing information-processing needs of business are converting computers into forms that increasingly resemble telecommunications networks, whereas the old telephone system has already become, in a very meaningful sense, a gigantic computer. But, more generally, reduced transportation costs, or the increasing prominence of industries in which such costs are relatively insignificant, such as electronics, make possible new divisions of labor. Such arrangements make it feasible for an industry to parcel out separate activities or components on a truly international basis. Thus, buyers of Boeing's forthcoming 767 will purchase an airframe the components of which have been manufactured in several countries on different continents; Boeing will happily deliver the aircraft equipped with either American (Pratt and Whitney or General Electric) or British (Rolls Royce) engines to suit the preference of the buyer. The European Airbus, on the other hand, comes equipped with American engines (General Electric).

The computer industry, at present, is still characterized by American dominance of world markets (about 80 percent of the total). However, with the growing Japanese domination of memory chips, there is apprehension of an eventual Japanese monopoly of that particular strategic component. A situation in which the United States dominates the brains of the computer, the logic circuits, and the Japanese the memory circuits, is perfectly feasible. Alternatively, although there is an increasing resort to Japanese components, American manufacturers continue to dominate the mainframe market. Finally, although Japanese performance in hardware continues to improve, American superiority in software appears to remain overwhelming. The Japanese failure to approach American sophistication in software accounts, in turn, for their inability to make larger inroads into the rapidly expanding minicomputer and personal computer markets, where software is especially crucial.

The purpose of these references is emphatically *not* to suggest that the present lines of demarcation between American and Japanese strengths will remain unaltered. That would be foolhardy. Rather, it is to suggest the possibility of entirely new patterns of specialization at the common technological frontier, as more and more nations reach that frontier. These new patterns may be such that it makes little sense to apply terms such as *technological leadership* or *technological gaps* to entire industries, much less entire countries. There may be persistent reasons for a country to retain technical and commercial superiority in specific segments of an industry while not dominating the entire industry. Moreover, it is entirely possible that countries that are generally regarded as far from the technological frontiers and at intermediate stages of industrialization – India, Brazil, Mexico – may establish niches for themselves in certain portions of high-technology industries.

In addition, we should recognize the limits of the explanatory power of the technological variable by itself. Success in high-technology markets is not reducible to mere technological capability. Such success is also a matter of organizational flexibility, managerial effectiveness, the social and economic rewards of risk taking, the efficiency of capital markets, and a number of other environmental factors that may affect the ability to convert technological opportunities into commercially successful innovations, whether these opportunities first emerge at home or abroad.

III

Finally, certain questions must be addressed concerning the interface between science and technology – and it is perhaps the intensity of activity at that interface that defines a high-technology industry in the first place. The long-term vitality of such industries is directly dependent upon the degree of success in achieving creative interactions between the two realms. Yet, we are far from having a really good understanding of the conditions that make for successful interactions, and how they can be encouraged.

To begin with, as suggested in earlier chapters, in crucial portions of high-technology industries, attempts to advance the state of the art are painstakingly slow and expensive because of the limited guidance from science. The development of new alloys with specific combinations of properties proceeds very slowly because there is still no good theoretical basis for predicting the behavior of new combinations of materials. Attempts to improve fuel efficiency are severely constrained by the limited scientific understanding of the combustion process. The development of synthetic fuels is seriously hampered at present by our ignorance of the molecular structure of coal and the relationship of that structure to its chemical properties. The designs of aircraft and steam turbines are both hampered by the lack of a good theory of turbulence. In the case of aircraft, wind tunnel tests still involve substantial margins of error in predicting actual flight performance. Indeed, the high development costs described in this paper are due precisely to the inability to draw more heavily upon a predictive science in determining the performance of new designs or materials. If science provided a better predictive basis for producing optimal designs, development costs (which constitute about two-thirds of total R&D expenditures in the United States) would be far lower.

On the other hand, as was argued in Chapter 7, scientific progress itself has become increasingly dependent upon technology. In advanced industrial societies, an important spur to scientific progress is given by the attempt to account for anomalous or unexpected observations or difficulties that arise in the productive process. In addition, technological progress in the design and construction of scientific instruments has enormously expanded the observational capabilities of science. Our understanding of the remarkably complex molecular structure of polymers, for example, would be far less in the absence of an array of twentieth-century scientific instruments–x-ray diffraction equipment, the ultracentrifuge, the electron microscope, the viscometer, and so on. To an increasing degree, the growing proximity between science and technology is proving to be a powerful stimulus to science, which in turn feeds back into the productive process.

This growing proximity between the scientific and technological realms identifies some important concerns that will require greater attention in the future. Decision makers in both the public and private sectors will need to determine how to improve organizational conditions and incentive structures at the science–technology interface. The ability to improve the functioning at that interface will undoubtedly be an important determinant of future leadership in high-technology industries. This is so not only for the reasons already indicated but because changes also appear to be occurring in science as well as technology. For example, there is evidence that forms of scientific knowledge most likely to be useful to high-technology industries must be pursued in an increasingly interdisciplinary fashion. The transistor was the work of physicists, chemists, and metallurgists. The scientific breakthrough leading to the discovery of DNA was the work of chemists, biologists, biochemists, and crystallographers. Historically, new disciplines such as biochemistry emerged when practitioners of separate disciplines – biology and chemistry – discovered an interesting range of problems at the boundaries of their respective disciplines. Unfortunately, such interdisciplinary research runs counter to the organizational arrangements, priorities, and incentive structures of the scientific professions, although admittedly that is more true in the academic than the industrial world. Nevertheless, there may be a high social payoff to enlarging the opportunities for at least some larger subgroups of the scientific community to define their activities with a stronger problem orientation rather than a discipline orientation, and in ways that will make it easier to undertake joint interdisciplinary research when it appears to be promising.

The most successful research institutions in private industry have already demonstrated that it is possible to conduct both fundamental and interdisciplinary research in a commercial, mission-oriented context. In addition, the United States has, in the past, shown great institutional creativity by inventing public-sector mechanisms for research in specific fields. The land grant colleges and agricultural experiment stations have played immensely important roles in exploiting the regionally diverse agricultural resources of the country; the National Institutes of Health have played a vital part in advancing the frontiers of medical research; and the National Advisory Committee on Aeronautics, the predecessor of the National Aeronautics and Space Administration, played a critical role in generating empirical data for the design of new aircraft that made an essential contribution to American worldwide leadership in the commercial aircraft industry. Devising additional ways of encouraging a closer and more creative interaction between science and technology seems to be both feasible and in close conformity with our pragmatic traditions.

Index